A Straightforward Guide
To
Health and Safety
Law

Samantha Walker

Straightforward Gui
www.straightforwardcc

Straightforward Guides

Samantha Walker has asserted the moral right to be identified as the author of this work.

© Samantha Walker 2017

978-1-84716-733-0

Printed by: 4edge Ltd www.4edge.co.uk
Cover design by Bookworks Islington

Contents

Chapter 1 - Company Structures and Employment
Status   10

Chapter 2 - Starting From the Beginning: The   14
Basics that Need to Be in Place

Chapter 3 - Health and Safety Law   23

Chapter 4 - From Asbestos to Zoonoses: General   56
Health and Safety Topics

Chapter 5 - Investigating and Reporting Accidents   148

Chapter 6 - Health and Safety Prosecutions   159

Help and Information

Index

****

Introduction

A few years ago, when I was working as a Government Health and Safety Inspector, the topic of my job would often provoke a reaction from people when they found out what I did for a living. The subject of health and safety is quite an interesting one for a lot of people, especially if they have had some involvement with it in their lives. Many people just wanted to hear any juicy stories about amputated arms and legs (thankfully there were not many of these to talk about!) and gruesome diseases that I had come across (again, I'm pleased to say that these were few and far between.) Other people asked for advice on a 'little problem' that they had at work, such as it being too hot/ too cold/too cramped or the fact that they had to use the factory next-doors' toilet as their employer refused to repair their own one. There were sometimes displays of scorn or indifference from the employee who thought that the Nanny State had, yet again, gone too far and that the government was spending too much time banning games of conkers than running the country. But one of the most regularly repeated comments by a variety of the people I came across, in social and work environments, was that 'health and safety' was boring and difficult to understand *as there was too much of it* and, as an employer or someone with health and safety responsibilities, they did not know where to start. There seemed to always be the argument that it consisted of a never-ending stream of laws and Regulations, followed by amendments, with too much time and money needed to put things in place, and too many conflicting opinions and no easy and credible source from which to get a coherent answer. With that mentality it is not really a great surprise that many companies want to bury their heads in the sand and hope that the spectrum of health and safety is something that they can put to one side until tomorrow, or next week, or next year, and it stays that way until unfortunately disaster strikes and they find themselves in court defending a breach of law that they never even knew existed. The reason for writing this book therefore, is to help those people to access the information that they need to put the necessary procedures and preventative measures in place to safeguard the safety and well being of themselves and their employees.

So why should we bother with health and safety? The latest figures show that in the year 2015/16 the number of work-related deaths in Great Britain was 144. 621,000 injuries were thought to have occurred at work. There were 30.4 million working days lost because of work-related illness and workplace injury, with 1.3 million working people suffering from a work-related illness.

Thousands of people still die each year from work-related diseases - the most common probably being that from exposure to asbestos. The Health and Safety Executive estimates on its website that there are between 3000 and 12000 deaths each year from occupational cancer, with some 4000 of those probably being asbestos related. That figure is almost sure to keep rising, with an estimated peak level projected for the years between 2011 and 2015. Less people under the age of 55 are dying from mesothelioma (a form of cancer affecting the lungs and other organs) these days which suggests that better controls have been in place in recent years, but the country is still suffering from the consequences of the (known and unknown) dangers of the past. Protection against common day hazards and hazardous materials is vital if we are to avoid these debilitating and painful causes of death for future generations. No deaths are acceptable - we go to work to earn money to live, not to lose our lives or become injured or un-well. It is therefore imperative that health and safety matters are taken seriously.

Another important point to mention is the cost of not complying with health and safety law, especially since the new sentencing guidelines came into force. For cases taken to court by the Health and Safety Executive in 2009/10, the average penalty per conviction was £15,817. In 2015/16, the average penalty was around £58,000 per case resulting in conviction. For a small company, this sort of loss can lead to severe financial hardship, especially if there are the prosecution's costs to pay on top of this, and compensation payouts to victims and/or their families.

Aside from all the doom and gloom, there are numerous benefits to be experienced by companies who take health and safety seriously and apply it as an active, integral part of their daily work-lives. A positive health and safety culture boosts moral and in turn makes employees feel valued and looked after, which helps to increase their general well-being and attendance at work. Having a good health and safety management system can reduce the amount of accidents/ injuries/near misses and reduce the number of lost working days. Employees may be fitter and more able to do their jobs, with a reduction in the length of time that they need to stay off work after an injury. A big bonus for some companies may be the money aspect –depending on your insurer there may be cheaper (or at least no increased) insurance renewals and a reduction in civil claims. Having a structured approach will enable you to document where you stand as an organisation and give you better data for auditing and benchmarking (against yourself or other companies.) It will also show that there is commitment coming from 'the top' and is therefore an assurance that management are taking health and safety seriously. And finally, anyone can check the Health and Safety Executive's website to see if you have any enforcement notices or prosecutions against you, so if you have a clean slate it can help you to get picked over other contractors when bidding for contracts, for example. Now doesn't this sound good?

Change can be difficult - but with the right approach it is possible to get even the most cynical of people on board. From my experience, the best thing to do is to say what you are going to do AND THEN DO IT.

People tend to become cynical and disbelieving if they have heard things before but have never seen them come to fruition. Measures such as giving employees enough information as to why the change in culture is necessary (and how it will benefit them and the organisation as a whole), and also letting them have input into how it should be implemented, along with clearly defined steps as to how it will be done and what is expected of them along the way, is a good place to start. Of course, if you are thinking of starting a business, or have yet to take on any employees, you can aim to get it right from the beginning.

A good thing to remember is that, once you have everything in place and up-and-running, it is basically a case of making sure that everything is working and that you review and update your policies and procedures as necessary. Of course, there will always be change in your working practices, equipment or people that you employ, as that is the nature of business, but you should have enough knowledge and confidence to not be flummoxed by this, and will instead know where to go to get the information that you need, and will understand what you need to do with that information to update your health and safety management system. Health and safety law is introduced in the UK in April and October of every year, so it is worth keeping an eye out to see what may affect you in the months leading up to these dates so that you can be ready. The list of contacts for further information at the back of this book can point you in the right direction for finding out more information.

The manufacturing base has declined significantly in this country in recent years, perhaps because it is cheaper to manufacture and transport from abroad than to do so in this country. Small to medium-sized enterprises (SMEs) have become more prevalent and have become a large part of industry in Great Britain. Amazingly, 99% of the businesses in the UK have fewer than 50 employees. There are several different calculations as to how many employees/amount of turnover constitutes an SME, but in this instance I have chosen 0-249 employees, although some definitions define one as having 1-49 employees. Perhaps due to the smaller numbers employed by some employers, less than 30 % of the workforce now belong to a Union - this is why it is necessary for companies to have nominated health and safety representatives from within the workforce to represent the views of employees and have help towards creating a good knowledge base of health and safety within the company. Health and safety is a changing environment- the risks seen these days are completely different in some industries to how they used to be years ago, in particular stress and musculoskeletal disorders (MSDs) are now playing a large part in the different risks to which workers are exposed to and therefore it is necessary for companies to keep pace with new hazards and the control measures available as they arise. It is often the case with SMEs, perhaps due to a lack of resources (personnel finances and time) that they do not have a person dedicated solely to delivering health and safety commitments, or access to specialist help to enable them to do so, or that they do not feel confident in tackling the relevant issues themselves. This book is for those categories of people or organisations, and I hope that by the end of it you will feel that, with a bit of work, it really is not that bad after all.

The aim of this book is to show how health and safety law affects employers, employees, self-employed, contractors, and in certain cases the public and those not employed by you, but who may be affected by your work activities, and what is required of you as an employer and/or employee as you may be both. Where I have specifically written 'you' please take this to mean you as the dutyholder, i.e. the person or organisation with health and safety responsibilities and overall duty of care for the workplace and the persons contained within it or affected by it. The basics contained within this book will, in most cases, apply to all companies and places of work which employ people and who have people on site. This book doesn't cover regulations that are specific to construction or farming, or law that is not common to most industries, as there would not be the space to include everything. A brief reference to the new Construction Regulations is made further on in the book. No one book of this size will tell you everything that you as a company or individual will need to know, as every company is different with a varying workplace and activities and new legislation is frequently introduced, but I hope that you will use the information contained within it as a useful building block for creating your own individual health and safety management system and culture, whether you read it from cover to cover or just go straight to a specific section should the need arise.

****

Chapter 1

Companies, Structures and Employment Status

Before you start doing anything you need to know what type of legal status applies to you as an individual or a company and those that undertake work for you on your behalf, as there are numerous different types of employment status such as sole trader, self-employed, partnership, limited company, employee, and this can affect your and other people's responsibilities under health and safety law.

You can get advice on company structures, as with all matters relating to law and business, from an accountant or solicitor as appropriate, on the specific legal structure that would suit you best for your company. There may well be pros and cons associated with the legal structure that you choose.

In brief, to be a partner, sole trader or a member of a limited liability partnership as an individual, you must be self-employed and register as such with H M Revenue and Customs –which means that any work you do for this particular business is done on a self-employed basis. As a sole trader you take the responsibilities for health and safety yourself; in a partnership the two or more persons involved would share these duties. Unlike a limited company a partnership has no legal existence distinct from the partners themselves. Limited companies exist in their own right, therefore the companies' finances are separate from the person's finances. Private and public limited companies must be registered at Companies House and must have a least one Director (two if a PLC) and a Company Secretary who may also be shareholders. However, from October 2008 it hasn't been be necessary for private limited companies to have a Company Secretary.

The name of a limited company must end with the word 'limited' or 'ltd'. A limited company can commit a health and safety breach and it will be the company, in general, who will be prosecuted if proceedings are taken against it. Directors, managers, employees and others can all be liable as individuals under health and safety law. Company Directors can be disqualified through the Courts under certain circumstances. A company also need not have any employees. If a company changes its name, this does not affect any legal proceedings made against it in the former name. Trading names have no legal status and the real name and a specified address (not a PO Box number) must be used on correspondence and letters. This information must also be displayed at every place of business where the company operates. A partnership cannot be prosecuted as it has no legal 'personality' in terms of criminal law- the individual partners are therefore potentially jointly liable for an offence. Partnerships shouldn't be confused with limited liability partnerships, though, which *do* have a legal personality. Limited liability partnerships must also register with Companies House. In certain circumstances, companies in insolvency procedures may also be liable for prosecution for a health and safety offence. Once a company dissolves it ceases to exist and therefore cannot be prosecuted for a health and safety offence.

So, once you are clear on the legal structure, you can then work out whether those that you give work to are direct employees or not, as more legislation may be applicable to direct employees although there are still many duties owed to the self-employed and contractors who may work for you. In terms of health and safety, employment status is not always clear-cut. Indicators of a contract of employment would include 1) whether the employer has the power to select and appoint the individual, and to dismiss or suspend them, along with paying their wages and holiday and sick pay, 2) whether the employer supplies the tools and equipment, and fixes the times and place of work, and 3) the type of contract - even a verbal contract can stand in a court of law. Payment by job, as opposed to being on the PAYE system, may point away from employment. Look at the terms of the contract that you have with workers - it is always best to seek advice on these issues, if you are not sure.

If you employ anyone you also need to have Employers' Liability Compulsory Insurance and this certificate must be displayed in your workplace (this can be done electronically) and must obviously be renewed when it runs out to ensure continuous cover. This is to insure against injury or liability for disease to employees arising from their employment. You can be fined if you do not hold a current Employers' Liability Insurance policy. If an employee is injured at work or even a former employee becomes ill as a result of their work whilst in your employment they may try to claim compensation if they believe that your company was responsible. The Employers' Liability Insurance that you take out will enable you to meet the costs of compensation. Many road accidents which occur whilst employees are working for you may be covered separately by your motor insurance. *Public Liability* insurance covers you for claims made against you by members of the public or other businesses but not by employees. Having Public Liability Insurance is generally voluntary, depending on your business. For Employers' Liability Compulsory Insurance you must use an authorised insurer (which you should check they are before you take out the policy.) The Financial Conduct Authority (FCA) maintains a register of authorised insurers which you can find at www.fca.gov.uk. You will have an agreement with your insurer as to the circumstances in which they will pay out. For example, the policy will cover the specific activities which relate to your business. They cannot refuse to pay purely because you have not met a legal requirement connected with the protection of an employee but this does not mean that you can forget about your legal responsibilities. If the insurer believes that you failed to meet your legal responsibilities for health and safety to your employee and this has led to the claim, they could then sue you to reclaim the loss of the money paid in compensation. Currently you must be insured for at least £5 million but most insurers offer cover for more than that. The Employers' Liability (Compulsory Insurance) Act exempts the following employers: health service bodies, most public organisations, and some family businesses (so if all of your employees are closely related to you then it may not be needed.)

Be aware that some people who you may think of as self-employed may be considered as employees for the purposes of the Employers' Liability Insurance and you would need to check this out with the insurer. It does depend on your contract with the 'employee' - normally if you deduct NI and Income tax from their wages then they will be classed as an employee, and if you supply most materials and equipment that they use, and if you control their working hours. You must tell your insurance company if you take on people who are unpaid, such as school children, or volunteers. Motor vehicle insurance will also be needed if you supply vehicles for people to use as part of their work. You must keep copies of the certificates of insurance for at least forty years as diseases can take a long time to surface. If you do not display a certificate of Employers' Liability Insurance or refuse to make it available you can be fined, and you can also be fined for any day on which you are without this insurance.

****

Chapter 2

Starting From the Beginning-The Basics That Need to Be in Place

Health and safety law has its origins in the Robens Report of 1972, which formed the basic idea for the Health and Safety at Work etc. Act 1974, which is the main piece of legislation governing health and safety law in the UK. This also introduced the idea of having an authoritative body, known as the Health and Safety Executive (HSE) which develops policy, secures compliance and enforces the law whilst aiming to protect people at work and ensure that all occupational risks are controlled properly.

Health and Safety law is enforced by Inspectors from the Health and Safety Executive or by Environmental Health Officers (EHOs) from the Local Authorities. EHOs have similar powers but they cover different types of premises to the HSE.

Like the HSE, Local Authorities enforce the laws as made by Government. There are over 400 Local Authorities with responsibilities for enforcement, so they make up a significant part of the role of the enforcer. The Local Authorities also cover food safety, and areas such as housing as well as some health and safety issues.

The HSE currently comes under the Department for Work and Pensions and is therefore accountable to Parliament for their activities and performance although they remain largely independent. Local Authorities are the principal enforcing authority in retail, sales, warehousing, catering, offices, hotels and the leisure industry, although responsibility can be transferred between HSE and Local Authorities by agreement. They also inspect each other, although I don't think this happens very regularly! HSE deals with factories, motor vehicle repair workshops, hospitals, schools, universities, offshore gas and oil installations, farms, the movement of dangerous goods, and construction, amongst other workplaces. There is often an overlap and you may end up being visited by both enforcers! HSE carries out scientific research in conjunction with other organisations, and consults industry and the Trade Unions via consultations and committees. The HSE also includes Occupational Health Inspectors, the Nuclear Safety Directorate, and the Hazardous Installations Directorate (who deal with sites who have large quantities of hazardous chemicals) and the Laboratories (HSL) which conduct experiments and analyse evidence.

Health and safety law applies to all businesses, no matter how small, and covers all types of employees - full and part time, temporary and permanent, the self-employed, those on work experience, apprentices, charity workers, mobile workers and home workers. You will also have responsibility for any temps and casual workers brought in from agencies. If you use temps or agency workers you must tell the employment agency who hires them about the risks to that worker's health and safety and the steps you have taken to control those risks, any necessary qualifications or skills needed to do the job, and any health surveillance that will need to be undertaken. The employment agency must then pass this information on to the worker in whatever form they feel appropriate but you must ensure that information has been received and understood so it is probably better for you to also give this information to them yourself. You need to ensure that they have understood what you have said and measures may need to be put in place if they have any special needs, especially if they do not understand the language and/or English is not their first language. Some leaflets produced by the HSE have been translated into other languages which you might find a useful aid. You have also got to protect the health and safety of the public from any risks caused by your work activities.

Health and safety Inspectors have the right to enter work premises without giving prior notice although often notice is given if it is deemed necessary. During preventative inspections they are more likely to turn up without warning to see the business as it usually operates, on a normal day. Inspectors have enforcement powers that they can use if health and safety standards are not found to be satisfactory. These include Improvement Notices, which require problems to be rectified within an agreed time period, and most often the Inspector will re-visit to check that the remedial action has been taken. Inspectors can also serve Prohibition Notices, with deferred or immediate effect, where there is deemed to be an activity or circumstance which involves, or will involve, a risk of serious injury. A Prohibition Notice means that until you have complied with the remedial action the activity will not be allowed to resume. Further information about Notices can be found in the section entitled 'When things go wrong.'

HSE inspectors also prosecute where serious, or consistent, failings have been found. Most cases are heard by Magistrates in England and Wales, who may for serious offences impose a maximum fine of £20,000. If cases go to Crown Court there is no limit on the fine which may be imposed. Manslaughter investigations reside with the Police, although the HSE/Local Authority may work in conjunction with them to assist with specialist knowledge or with the gathering of evidence. In Scotland most cases are taken in the Sheriff's Court either on summary procedure or on indictment procedure before a jury. There are a number of differences to be found in Scotland, namely in that Scottish HSE Inspectors do not prosecute as they do in England and Wales; instead the HSE reports to the Crown Office and Procurator Fiscal Services who decide whether or not to instigate proceedings. The reports put forward by the HSE to prosecute can be rejected.

The HSE also has Health and Safety Awareness Officers (HSAO's) that undertake certain roles. Unlike Inspectors they do not have powers to enter your premises without your agreement and can only enter with your consent. The limited powers that they do have are to enforce certain aspects of the Employers' Liability Compulsory Insurance legislation. They are trained staff whose role is to support the HSE's regulatory work. They aim to provide information and will often bring free literature and leaflets with them to promote health and safety awareness. They are also there to collect information about your business which can be used by the HSE in further contact with you and to gain information about your activities and numbers of staff that can then be used should an inspector visit your business. They are not Health and Safety Inspectors and they will not give technical advice or opinions on how you are managing health and safety or your compliance with the law. They may talk to your health and safety representative, as will an Inspector, and will send a copy of any correspondence that they send to you to the representative as well. If an HSAO thinks that you need more information than they can give you, or if they think there are health and safety issues within the workplace, they may suggest that an inspector also undertakes to visit in the near future. Health and safety representatives should be given the opportunity to speak privately to the HSAO and indeed on a visit an Inspector may often request that they have a word alone with anyone that wishes to speak to them privately.

A large part of an Inspectors role is to undertake inspections (funnily enough). HSE Inspectors tend to have a specified area (such as a county) in which they work and they try to have an up-to-date list of premises in that area that they try to visit. Many inspections also take place in response to a complaint. A company with a history of non-compliance may be visited more frequently than other workplaces to ensure that workers are protected.

Starting up

Starting up a business or starting to implement health and safety law in your business can seem at first quite a laborious and confusing task. Where do you start? There are certain components that you need to have in place as a small to medium enterprise, or indeed any business, to ensure that you are complying with health and safety law.

Firstly, you must employ one or more Competent persons to provide health and safety assistance. The competent person should ideally be appointed from within your workplace, as they are technically always on hand, or it can be yourself as the duty holder if you think you have enough knowledge or time or you have few employees. Alternatively you can use external services such as a consultancy. The employee(s) will need to be given time, resources and training to enable them to provide this service and this needs to be provided by the company. External services can also help with health surveillance, worker rehabilitation and other services requiring a level of expertise.

You also need to say how you are going to manage health and safety in your business and this information needs to be written down if you have more than 5 employees. This is called your **Health and Safety Policy**. This basically shows who in the company does what regarding health and safety, and when and how they do it, and how you intend to manage health and safety as an organisation. Risk assessment (covered in the section 'Management of Health and Safety Regulations') forms a part of your health and safety policy as this is how you will identify the hazards, and therefore any risks, to those in your workplace, along with any control measures that are needed to reduce those risks. The Health and Safety policy should include a statement of 'general policy', which should be signed by the Director(s) and whereby you declare that you will, amongst other things:

- Maintain adequate control of risks arising from work activities,
- Provide safe equipment and workplaces,
- Ensure safe use and storage of substances,
- Provide adequate information, instruction and training to employees to enable them to do their job safety,

18

- Consult with employees on health and safety issues,
- Prevent accidents and ill health and investigate accidents as necessary (as well as dangerous occurrences),
- Maintain healthy and safe working conditions, and
- Review and revise your policy as and when necessary.

The policy should also detail the overall responsibilities for health and safety and who is accountable for them, the person(s) that are responsible for putting the policy into practice and for reviewing it, the roles and names of employee representatives (union or work-appointed by employees) and how consultation and the giving of information takes place.

The specifics that you might also like to look at include:

- Who will ensure that inspection and maintenance procedures are drawn up and undertaken?
- Who will look to see what new legislation is coming out that might affect the company?
- Who will ensure that the procurement of new plant and equipment is checked to ensure it satisfies health and safety requirements?
- Who will do risk assessments and COSHH assessments and also review them?
- How will info for new employees be given, for example in inductions?
- Who will organise monitoring and health surveillance, if necessary?
- Who will organise first aiders and ensure their qualifications are up-to-date?
- Who will look after emergency procedures and fire, such as doing a fire risk assessment and reviewing it? Who will test the alarms, and

19

ensure that fire extinguishers are maintained and escape routes are clear at all times?

- And any other specifics which relate to your business.

Responsibilities of employees

It is important to name the responsibilities of employees, too (such as to not interfere with anything that has been put into place for the sake of protecting their health and safety; to co-operate, so for example to wear hearing protection if it has been found necessary to do so and it is therefore provided for a purpose, and to report any health and safety concerns to an appropriate person.) Discuss the health and safety policy with employees and safety representatives as they may want some input on how things are managed- they are often the best people to talk to as they know how things operate on a day-to-day basis and can let you know of anything that may not work. Review the health and safety policy periodically - i.e. at least once a year, and definitely when any processes or activities change, or the workplace changes (i.e. the layout of the building or new machinery). The health and safety policy and any risk assessments you do only need to be recorded if there are five or more employees within the company.

Information

There is also a requirement to provide sufficient information for employees on health and safety matters- you need to have a health and safety law poster if you employ anyone, alternatively if you do not think that everyone could have access to the poster because they work in different places or are based elsewhere, then you can give them copies of the same information contained within the poster. Employers have a legal duty to display the information contained within the poster (available from HSE books and some bookshops) in a prominent position in each workplace. There is a new health and safety law poster and leaflet that has been released by the Health and Safety Executive, which replaced the old format by 2014 at the latest.

You must also provide information on factors which may affect their health and safety at work. All of this information can go on a notice board along with other such information such as fire procedures, escape routes, first aiders and any other company communications relating to health and safety.

Accidents

As part of your management of health and safety, you will also need to record any accidents which your employees and others may have at work. The 'official' way to do this is to use an accident book which enables you to record the details of the accident and who it happened to. Since 2003, all accident books must now comply with the Data Protection Act which means that they must have a way of preventing anyone reading the accident book from seeing the personal details of the persons listed in the book, such as information on where they live. Usually this means that the personal details section is torn out and kept in the employee's personnel file, whilst leaving a corresponding page in the accident book that just gives brief details of the accident. Accident books can be purchased from HSE books and most good bookshops.

Environmental impact

It is also important to look at the environmental impact that may occur when setting up in business, especially these days when people and industry are trying to reduce their carbon footprints – consider the building materials, any waste produced, and whether the company will be using its resources effectively. Look also at how you will dispose of solid/non solid and toxic waste. What chemicals do you use - could they leak into the watercourses or affect flora, fauna and wildlife? What do you have around you in terms of wildlife and ecological systems? Consider any noise that you make and the effect that it may have on the surrounding environment.

Where are you now?

Taking things forward

In analysing your company's current health and safety performance, and thereby learning about the areas you need to improve on, it is useful to think about the following questions and the (truthful!) answers that you find.

1) What have we been getting right, and where have we fallen short? What reasons are there for this?

2) How do we fare in respect of our original health and safety aims and objectives? Have we got better or worse?

3) Are we adequately controlling hazards and risks? How do we know?

4) Have we got it right in terms of the time and resources we put into health and safety management? Is our health and safety management system effective and efficient, and working across **all** areas of the company/ organisation?

5) Do we have a positive health and safety culture, reinforced by management commitment and action, as well as that of employees?

6) How do we compare with others? Do we have ways of sharing information with, and learning from, other similar companies?

****

Chapter 3

Health and Safety Law

Law

There are two major types of legislation concerning health and safety at work. The first type are the Acts of parliament, of which the most important is The Health and Safety at Work Act 1974 as this forms the basic groundwork for all of the other legislation relating to occupational health and safety. The second type of legislation includes Regulations, also known as 'Statutory Instruments.' These are normally concerned with one aspect of health and safety, such as work at height, or asbestos, and are often accompanied by an ACOP – An Approved Code of Practice. Much of current law tells you what you need to do, but not necessarily how you should go about doing it. This is where ACOPs come in handy -they give advice on how to comply with the legislation and often cite examples and solutions to generic problems.

Regulations need to be laid before Parliament, and become law 21 days after being submitted unless objections are made. ACOPs are approved by HSE and do not require agreement from parliament. ACOPs have a special authority in that whilst they are not 'law', as such, they can be used in court against a company or individual unless it can be shown that the recommendations found in the ACOP have been complied with in an alternative and equally effective other way. So you must be able to show that the alternative action you took still met the standards sufficiently. ACOPS are useful in that they provide flexibility in complying with the law. 'Guidance' is also produced by the HSE and other authorities, often in the form of free leaflets. This guidance is not compulsory and employers are allowed to take any other action as they see fit, as guidance is just that - guidance. However, following the actions and methods described within these guides is often enough to show compliance at a basic level.

Much of health and safety law that has recently been implemented has been on the back of European Directives, or standards which derived from good practice. British Standards are also important as guidance- they help manufacturers in the design and build of products and enable a consistent approach to achieving acceptable standards within industry. A recent report by Professor Lofstedt, who was asked by the Government to look into the UK's system of health and safety legislation, had the objective of aiming to streamline the body of legislation and to improve the way it is enforced by the HSE and Local Authorities. As a result of the Professor's findings, many Regulations have been removed or changed and it is likely that more will follow in the future.

Health and safety law comes under criminal law, as opposed to civil law, and it is therefore a criminal offence to commit a breach of health and safety law.

'So Far as is Reasonably Practical'

An important phrase that you will see time and time again is that of 'so far as is reasonably practicable' or 'SFAIRP' which is used often in health and safety law – it means that were it reasonably practicable for you to take risk control measures then you must do so, up to the point where the implementation of measures would be disproportionate to the degree of risk and therefore the risk is minimal in relation to the sacrifice that is undertaken to control it. Basically it can be seen as a balance of scales, whereby if you have risk on one side, and elements such as time, effort or money on the other, then if the risk is far 'heavier' than the measures it would take to reduce that risk, then it is seen as reasonably practical that you take those measures to reduce the risk, as the risk has far outweighed the time and effort etc that it would take you to put it right. If a breach of law is brought to court, the onus is on the accused to prove that it was not practicable to do more than was actually done to satisfy the requirement, and/or that there was no other better means with which to do it. A hard one to prove! Similar to SFARP is ALARP – 'As low as reasonably practicable' - which deals with whether the risk has been reduced to what might be deemed an 'acceptable' level.

The Health and Safety at Work etc. Act 1974

What it says

This Act of Parliament is the main piece of legislation concerning the general health, safety and welfare of persons at work, and any others that may be affected in connection with the work activities of a company. It is commonly known as 'The Act.'

I have included the most important sections of the Act and what they relate to here, as this is the piece of legislation you are most likely to come across in an enforcement notice as it is really the blanket coverall for all offences, if a more suitable one cannot be found under a specific regulation. The Act is split into different 'sections' as opposed to 'regulations.'

Section 2 declares that it is every employer's duty to protect the health, safety and welfare at work of all his or her employees, as far as is reasonably practical. This should include having in place safe systems of work, safe equipment and tools, arrangements for the safe use, storage and handling of substances and materials, information and training for employees, sufficient maintenance of the workplace to make sure it safe to work in (with safe ways of entering and exiting it, and with adequate facilities provided), and a safe working environment within which to work. There is also a requirement to also consult with your employees regarding health and safety arrangements to give them opportunities to help devise measures to ensure their own health and safety at work.

If it is requested by safety representatives then you should also set up a health and safety committee which should meet regularly and discuss relevant issues.

Section 3 states that your work activities should be undertaken in such a way that, so far as reasonably practicable, persons not in your employment will not have their health and safety affected by being exposed to the risks found within your undertaking. You therefore have the same duty towards non-employees as you have towards employees. (This also applies to the self-employed.) You will need to tell these people, in both cases, any information you have relating to the risks to their health and safety.

Section 4 states that if you provide premises for persons to work in who are not your employees, and you have (to any extent) some control over any part of those premises, or over the plant contained within it, that you must ensure that the premises and that which you provide within it are safe and without risks to health and safety. This will apply to you even if you just have obligations to repair or maintain the premises.

Section 6 applies to manufacturers, importers and suppliers of anything to be used for work purposes, to ensure that any goods, materials, substances and other such items are designed and made so as to be safe and without risks to health and safety. This may require the arrangement of testing and examinations to determine this. You must also pass on any health and safety information to those that receive the items for use at work, and any further information that becomes available that may bring about risks to health and safety. This also applies to fairground equipment.

If you are involved in this area then you will especially need to read the Act in full.

Section 7 discusses the general duties of employees at work which include the requirement to take reasonable care of the health and safety of themselves and others who may be affected by their actions, and to co-operate with their employer with regards to the employer's duty concerning the protection of health and safety at work.

Section 8 applies to anyone at work, and states that no-one shall deliberately interfere with or misuse anything provided by their employer in terms of health, safety or welfare, for example items such as guarding or hearing protection.

Section 9 states that no employer can charge an employee money for anything provided or done within the workplace with regards to health and safety. For example, if your risk assessment has deemed it necessary for you to provide safety footwear, you must provide these free of charge. Some companies offer a range of necessary safety equipment, with those employees wishing to have an alternative design to the standard one offered often making up the difference in cost.

Section 20 details the powers of Health and Safety Inspectors, which include being able to enter any place of work at any reasonable time if deemed necessary to do so and to take a Police Officer if he or she thinks they may be obstructed from doing so; to direct that premises or items be left undisturbed for as long as necessary; to take samples, photographs and measurements of items; to ask that items if deemed dangerous be removed or dismantled; to take possession of items; to have questions answered, and to take copies of documents or other artefacts. The last power is the catchall power, which gives the Inspector 'any other power which is necessary in carrying out his role' which is rarely used and is very hard to determine!

Section 25 details the power of the Inspector to deal with causes of imminent danger and to seize anything which may be a likely cause of imminent danger or serious personal injury.

Section 33 details the offences which can be committed which include contravening any part of the Act or any Regulations or any requirement put on you by an Inspector or by a Notice (Improvement or Prohibition); to obstruct an Inspector from doing his duties; to make false entries on documents or books; or to intend to deceive or to pretend to be an

Inspector. On conviction, many of these offences carry a fine of up to £20,000, and/or a prison sentence in the Magistrates Court.

Section 37 deals with offences made by Directors, Managers and similar officers of the body corporate associated with the consent, connivance or neglect on their part that contributed to the offence, and therefore proceedings can be taken against them as well as the company as a whole.

Other parts of the Act talk about the functions, powers and appointments of the HSE which are not necessarily relevant to this book.

What do I need to do?

As discussed in the Starting up section, you should already have prepared a health and safety policy. The information contained within your health and safety policy will come into play, here, as you will already have identified how you will achieve compliance with the points mentioned in Section 2 above. You should set up a health and safety committee if requested to do so.

You should include those not in your employment in your risk assessment. Look at contractors, self-employed, people on work-experience, children, persons living nearby, and neighbouring companies. Look to reduce the risks to these people as much as possible, and tell them about the risks as appropriate. Look at any premises that you control or have control duties and ensure they present no risks to health and safety, or that the risks are suitably controlled.

You should give instruction and training to employees about their duty to look after the health and safety of themselves and others, and to co-operate with their employer. Let employees know if anything changes within the health and safety policy or the risk assessments.

Don't let anyone interfere with any control measures you have put in place.

It goes without saying — but don't try to pull the wool over an Inspector's eyes by trying to hide something. And don't try to impersonate an inspector. I haven't heard of this happening before (why on earth would someone want to do that anyway?) but it carries a hefty fine.

The 'Six Pack'

This well-known group of Regulations came about after a European Directive was made to try to get conformity on basic laws throughout the European Community and to ensure basic levels of health and safety standards throughout. They are called the 'six pack' as there are six of them, and they come as a collective bunch of Regulations. They cover some of the big issues in health and safety, such as risk assessments and health and safety management, welfare and safety at work, work equipment and manual handling. They came into force in 1992 but most of them have had amendments or several amendments to them over the years, which is why they are not all dated from 1992.

Management of Health and Safety at Work Regulations 1999 (as amended)

This is quite possibly the most important piece of legislation, second to the Health and Safety at Work Act, as these Regulations include the need to undertake a risk assessment of all work activities which underlies the fundamental basic principle of identifying and controlling risk. They are also commonly known as the 'Management Regulations.'

Some people say that health and safety really boils down to common sense - I agree with that partially, but not in the sense that you cannot deal with the hazard if you don't know the nature of that hazard, and you can't control the risk until you know what the hazard is. So common sense may have a place in acting sensibly and not doing anything stupid, but that won't really get you anywhere in terms of managing health and safety.

The findings of many risk assessments done in most work places will contribute much more to a workplace than a common sense approach will, as the latter takes no steps to identify the hazards in the first place.

So what is the difference between a hazard and a risk? A hazard is something that can cause you harm, such as falling from a ladder, or being hit by a falling object. The risk is the likelihood (and sometimes severity) of that hazard actually causing the harm. The risk can be categorised as a probability (there is a ninety percent chance of Lucy falling off the roof) or qualitatively (the risk of falling off the cliff is unlikely/insignificant/very likely).

In turn, a control measure is something that you put in place, like a guard or a written safety procedure, to lower the risk and to make it less likely that someone will be harmed by a hazard.

Risk assessments only need to be written down if you have five employees or more - but in the event of an accident, how do you prove that you have done them? Ok, you may have told Bob the welder the findings of your risk assessment of the workplace and his activity, but if an Inspector asked Bob, would he be able to recall all of the points that you mentioned, especially if nothing has really changed and therefore you haven't had to update it?

It is always best to write it down, in my opinion, and then you can prove that they have been done, and also show that you have reviewed them and updated them if necessary over time. Risk assessments are done so that you can tell anyone who may be affected by the hazards in the workplace what those hazards are, and what you are doing to control them.

So, the Regulations say that, as an employer or a self-employed person, you must do a 'suitable and sufficient' assessment of the risks found within your workplace and its associated activities, to protect your employees and others from those risks that they may be exposed to, or that they may be affected by (in the case of those persons who are not your employees).You also need to regularly review any assessments that you have done, and certainly after an accident or work-related disease occurs, or if your processes, activities, workforce, equipment or premises have changed, as new hazards may therefore be present. Special risk assessments also need to be done for those individuals requiring it by law, such as young persons and new and expectant mothers, and for lone workers and those who work from home.

Qu. But why should we do risk assessments? We know our business well already.

There is one answer to this. How do you know if something is wrong until you look at it and assess the situation properly? Once you identify the hazards in a particular setting, you can then see what control measures should be put in place to reduce the risk completely or to an acceptable level (remembering that the action taken should be justifiably proportionate to the risk.)

Your risk assessment must cover all work activities, even non-routine ones. Use a systematic approach and break the workplace down into areas (such as warehouse, stores, offices) or look at individual risk assessments for different activities (such as cleaning, hot work, manufacturing) or by subject area (such as machinery, work at height, manual handling.) As long as you make sure you cover every work activity you won't go far wrong.

Qu. Who can do risk assessments?

The answer to this is that anyone can if they are competent to do so, and have the necessary knowledge of the workplace, the work activities and processes and the hazards that arise from them. They also need to have access to a copy of the Regulations to ensure that all the relevant points are complied with. There are many training courses available, some of the providers of which may even come to your workplace to give the training. The contacts list at the back of this book may point you in the right direction should you wish to find out about training courses for risk assessments.

Doing a Risk Assessment

There are many different ways of doing risk assessments, some of which can be very complicated (often unnecessarily so.) Below is a simple way of ensuring that you comply with the regulations by identifying the key aspects of what is required.

What to look for:

- What are the hazards? How could someone be hurt?

Remember, a hazard is something that can cause someone harm. Be as specific as you can – remember that the risk assessment or at least the significant findings of it will be shared with your employees and other, and just writing 'chemicals' doesn't really give any useful information to anyone. What about chemicals? Are they just going to harm you whilst sitting there in their tins, or is it when you open the tins and the vapours get into the air that eye and throat problems could occur? Are they likely to be splashed into a person's eye when they are decanting the chemical into another container? Be specific, and remember that there may be more than one unwanted outcome associated with each hazard.

- What is the risk (the likelihood of someone being hurt?)

How likely is it that someone will be hurt? You can use a formula for working out this answer, for example one that might give a low, medium or high risk reading, many of which can be found from health and safety publications. Alternatively, if you feel you have enough knowledge, competency and understanding of the hazards you may feel that you can specify the risk adequately without using a formula. You might also like to address this question after you have looked at what is in place to control the risk at present, as the risk may then be perceived as being lower than at first thought.

- Who exactly could be harmed?

Look at who is in the immediate vicinity, such as employees, visitors, contractors and so on, and then look at who might be further afield such as neighbouring companies, members of the public. Unless you want to, there is no need to specify individual names, especially if you have a lot of employees. (That would need a very large piece of paper, and would be very time consuming!)

Also, if not who, then *what* could be harmed? You could also look at damage to the environment, and perhaps property if appropriate.

-What is in place to control the risk at present?

What has already been put in place to control the risk? Look at all of the control measures that already exist, including measures such as guarding, ventilation, training, safe working procedures, personal protective equipment, workplace inspections, supervision, even safety notices placed on machinery. Work through systematically and you may find that there is more in place than you originally thought. Of course, it must be effective and be working to be classed as a control measure.

- What more could be done to control the risk?

After looking at what is in place already, look at what else needs to be done to lower the risk.

There is a suggested hierarchy of measures to follow to reduce risks which is to: eliminate the risk by getting rid of the hazard, substitute the material/ machine etc for something less hazardous, segregate the area containing the hazards so that it is not easily accessible, control the hazard at source by using guarding, ventilation etc, and use Personal Protective Equipment (PPE) such as hard hats, ear defenders etc to protect the individual. Normally the best measures to take are those which protect people as a whole, not just individuals. It is then necessary to look at the following two questions:

- Who is going to address the further actions needed, and by when?

There is often something else that can be done to reduce the risk, and therefore to ensure that it is done it is good to make a plan and a timeline for completion, especially if the control measure may take some time to implement. For example, if a machine guard is required, but it needs to come from abroad and that may take two months for it to be manufactured and made, then assign responsibility for chasing the guard and ensuring it is in place to a specific person or department. That way, things are more likely to get completed.

- When is the risk assessment to be reviewed, if not before?

You need to review risk assessments periodically, and as soon as there are any significant changes. You could mark on the risk assessment the date that it was done, and who did it, and then perhaps indicate on it that it should be re-visited before a certain time period has lapsed.

How you could record the risk assessment:

Hazards	Risk	Who could be harmed, and in what way?	Control measures in place	What more needs to be done
e.g. Falling from height when working on roof	Medium	Employees, contractors	Harnesses, safe working procedures, training for employees, policy of not walking on roof	Use mobile work platform to remove the need to be on the roof

A more advanced way of looking at the risk would be to include the severity (outcome) of the risk (perhaps using factors such as *nil or insignificant injury, minor injury, major injury, fatality*, for example), along with the likelihood of that risk occurring (perhaps *not likely, likely* or *very likely*, as an example) and then use the likelihood and the severity indicators to give an alternative risk rating. A *minor injury*, with a *low* likelihood would lead to a lower priority for action than a *major injury* with a *very likely* likelihood. Doing things this way can therefore help to prioritise actions for further control measures. Finally, look at how are you going to record the risk assessment – will it be held on a computer, or will there be paper copies, or both? Which is easiest way for it to be accessed and updated? How are you going to relate the significant findings to employees and others who may be affected?

Are any additional assessments required, such as those which require it under specific legislation such as hazardous substances (COSHH) and dangerous substances (DSEAR) or for individuals with special needs (new and expectant mothers, for example)?

The Management Regulations also cover these areas listed below:

Health and Safety Arrangements

There must be suitable and adequate health and safety arrangements in place, which means that there must be procedures in place to protect health and safety. Risk assessments obviously play a large part in this, along with the health and safety policy which identifies personnel with specific responsibilities to make sure they are carried out. All arrangements should involve employees who should be consulted on how things are done in the workplace. Your arrangements should include monitoring of the workplace – perhaps by doing planned and un-planned inspections, with a regular presence of management 'on the job' to ensure that control measures are in place, that they are working and that they are used. There should also be a commitment to investigating accidents and ill-health.

Provide Health Surveillance

The need for health surveillance for employees exposed to certain substances during specific work activities is implemented under The Control of Substances Hazardous to Health Regulations (COSHH) but requirements for providing health surveillance will usually be identified via a risk assessment. Whilst doing the risk assessment the assessor should pay attention to any identifiable diseases that may arise from the work activity or environment, and any likelihood that a disease or condition may occur.

Competent persons

The law says that you must appoint one or more competent persons to help you to comply with health and safety law, which would preferably be an employee as opposed to an outside source, as those working for you will have inside direct knowledge of processes and activities. You might however consider having a mixture of both. Consultants can be useful but do look around for one that comes recommended. Make sure that you know exactly what you will be getting for your money and make sure this is suitable for your needs. Do be wary of having someone who doesn't know your business well to do your risk assessments for you, as the duty holder is still responsible for the risk assessments that are completed for him. Consultants can be a good source of information but do check their level of training and experience. Ask if they belong to a professional body or organisation, and if they have experience of the type of work that you do. Will they help you if you happen to be investigated by the enforcing authorities, or are prosecuted? Is legal cover included as part of your contract?

Before you use the services of an outside consultant, take a few minutes to write down what you expect from them, how you want them to go about delivering their services, the outcome that is required, how they must feed back to you (by reports, verbal advice etc), your main point of contact within the company for them to deal with, and when you want the work done by. If you are not happy with the answers given to your requirements, then have another look around. The competent person(s) should also consult with the safety representatives for the company. They must also be given sufficient information and time to perform their role.

Procedures for Serious and imminent danger

You should have evacuation procedures in place in case of danger. Inform all employees as to the possible nature of that danger, and the extent of harm that could occur. The information required for this will likely be produced from the outcome of any relevant risk assessment completed. Practice runs should happen, such as those for fire evacuations. There should be set procedures to follow, and a list of the competent persons required to perform any specific duties. Telephone numbers for the emergency and rescue sources should be to hand, along with the procedure to be followed in the event of an emergency.

Give adequate information to other peoples employees

Ensure that you give enough information to non-employees working in your undertaking (and self-employed persons doing the same). If you give them the same information as you do your own employees then there can be no discrepancies. All employees and non-employees should have an induction covering the hazards and procedures of the workplace and the work activities that they are involved in.

Health and Safety Training

Health and safety training should be provided for employees on commencing employment, and throughout their employment, specific to their health and safety at work and the hazards that they may come across.

Qu. I run a small electrical firm and have employees working in other people's homes and businesses - how can I assess the risks to their health and safety if I don't know what hazards they might come across?

You can complete a generic risk assessment covering broad risks and common hazards. Workers (or supervisors) on site can then be given training and acquire the competency to do on-site risk assessments specific to that job and environment. Every risk must be assessed and sometimes that can only be done by being there.

Q. My employees and I sometimes work out of an office that belongs to another employer, who runs his company from there too - whose responsibility is it to do risk assessments?

You should follow their arrangements for on site, but you will need to do your own risk assessments for your company as you are responsible for completing a suitable and sufficient assessment of the risks to your employees and others from your work activities. You will doubtlessly also be doing different work activities to the other company. You will need to collaborate and co-operate with the other company on site and share any significant findings.

Workplace (Health, Safety and Welfare) Regulations 1992

These Regulations exist to ensure that dutyholders look after the health, safety and welfare of employees at work, and they also apply to persons other than employers who have control over any part of a workplace.

Maintenance of the workplace and equipment

All plant, machinery, tools, equipment and areas of the workplace must be suitably maintained and kept in a good, clean condition, and there must be a suitable system of maintenance in place for plant and equipment. This system should be clear to all, and those in charge of the items must be aware of the type of maintenance and the frequency at which it must occur, and who is to do it. Regular tests and inspections should also take place, if appropriate and applicable, to ensure that it is safe and fit for use. All defects must be rectified and records of maintenance, inspection and testing kept somewhere safe.

Comfort

You need to provide a well-ventilated workplace (with fresh or purified air.) It is often enough to ensure this happens via the use of windows, or by mechanical ventilation systems. The temperature should be relatively consistent and at least 16C (or 13C if excessive physical work is involved in work activities) This temperature level applies to areas which are normally used for work for longer than short periods and will not necessarily apply to areas that do not get used. Work in cold rooms and the like will need to have extra clothing provided and measures to protect against the cold. You should make a thermometer available, perhaps by putting it on the wall in a main work area so that employees can see what the temperature is.

There should be enough space (and therefore floor space) and height to work in comfortably – the Regulations state that there should be 11 cubic metres per person as a bare minimum. You should also take into account any furniture, walls and stairs that may impact upon this space. Staff must be able to move around freely and will therefore need enough unoccupied space to achieve this. Working in cramped spaces can lead to ill-health such as back and neck problems, and can make people feel stressed.

There should, ideally, be separate toilets for men and women unless there are lockable doors from the inside. You must provide suitable washing facilities, with clean, hot (or at least warm) and cold water. There should also be suitable lockers or designated areas for employee's clothes etc, and facilities for changing, especially if uniforms or specialist clothing is to be worn for the purposes of work. There must be suitable rest facilities provided for employees to take breaks in, and somewhere for pregnant and nursing women to rest if necessary.

Safety

There must be suitable and sufficient lighting in the workplace (preferably by natural light) to avoid eyestrain, and to ensure that workers can see what they are doing. A lack of proper lighting can often cause an accident. Do watch out for fire hazards when stacking anything near lights, especially if the lights can get very hot over time. Ensure there is a substantial gap at all times, between the light and anything flammable. Floors must not be slippery, or be dangerous for other reasons such as that which contains cracks and potholes. Watch out for slopes outside, especially if it is snowing or raining or the ground is covered in wet leaves as this can make a slope dangerous for pedestrians and vehicles.

Look to see whether you need to install handrails on any steps or stairs. Ensure that traffic routes are free from obstruction. You must prevent people falling, or being struck by a falling object – look at installing guardrails or fences on pits, and ensure that tanks and containers are covered, to prevent people falling into dangerous substances. Install toe boards and guardrails if people are working at height to ensure that nothing can be dropped or kicked off and thereby hurt someone below. Make sure that no one can fall out of a window. Windows should be made of a suitable material which complies with the relevant British Standard. If you clean the windows in-house, work out how to clean them safely.

Pedestrians and traffic must be able to operate safely, i.e. separately and without coming into contact with each other.

A supply of clean drinking water must be available for all persons, with cups provided.

Qu. My employees sometimes work in other people's premises – how do I know that they will have access to toilets, rest areas and clean drinking water?

You should take steps to ensure that the above are available for your employees, by consulting the owner of the premises to see that they will be available. In some circumstance, such as on construction and mobile sites, it may be necessary to provide temporary facilities or an alterative means of provision.

Health and Safety (Display Screen Equipment) Regulations 1992 (amended 2002)

There is often some confusion between the definition of VDUs (Visual Display Units) and DSE (Display Screen Equipment). In actual fact, they are pretty much one and the same thing and for most of us the only time we will need to comply with these regulations is during our use of computers and associated equipment.

DSE is defined in the Regulations as encompassing any display screen, such as that of a computer. It does not apply to televisions, and it does not apply to equipment on a means of transport, such as satellite navigation systems. A Workstation in the Regulations is defined as involving DSE and other items such as desks, telephones, printers and the environment immediately around the DSE. As such, if you use a computer it will be the 'unit' within which you work.

A DSE 'user' is an employee or self-employed person who uses the DSE as a 'significant part' of their normal work. This applies to home workers too. A 'user' can also be further defined as someone who needs to use the DSE to perform his or her job and has to use it continuously for an hour or more at a time and does so perhaps more or less daily. This means that not everyone who works with a computer is defined as a 'user'. A data entry person, or a secretary would normally be deemed to be a 'user', for example. An IT Maintenance Engineer may not be covered under the Regulations, therefore, although he is involved with DSE every day. He would not be classed as a 'user' if he was just fixing it or putting it back together.

Doing a risk assessment will help to identify any 'users' that you have, and the associated hazards and control measures. You must ensure that you look at temps too, and any self-employed workers and contractors who work for you. You can always do a risk assessment for everyone who uses a computer if you are not sure if they are to be classed as a 'user' or not.

DSE assessment

The effects most often associated with DSE use are mostly musculoskeletal problems such as carpal tunnel syndrome, Repetitive Strain Injury (RSI), back, neck and shoulder problems ,eyestrain and stress. Looking at people individually and observing their posture and use of the equipment can often immediately highlight any problems as well as looking at the way the task is carried out.

Ensure that all assessments are done by those who know the requirements of the Regulations, and that they are done in conjunction with the 'user' as their input is required as to their own personal circumstances when using DSE. Some companies have nominated trained DSE assessors. You should aim to record the assessments, especially if there are any significant findings. As with all risk assessments, they should be reviewed periodically, to see if there are any problems, or changes such as new computer packages, desks etc, or there is a difference in the time spent at the DSE/workstation. The individual assessment could cover, but not be limited to, the following:

-Is there any flickering on the screen, or glare or excessive brightness?
-Is the screen height /unit/chair height /desk height adjustable?
-Is the lighting and temperature at a comfortable level?
-Is everything close by on the desk or does the user need to stretch to reach things?
-Do they need a foot or wrist rest?
-Is the user able to rest their wrists/arms in front of keyboard?
-Are breaks taken at regular intervals?
-Is any noise or vibration emitted from the equipment?
-Does the user find it easy to use the computer software and to set up and adjust the equipment?
-Is there enough space under and around desk?
-Are there any problems that the user has, or any health conditions that may be related to use of the DSE?

Any significant health problems that are identified may need to be referred to a doctor or your occupational health provider. You could devise a checklist to ensure that all points are covered for each user.

Regular breaks are needed when using DSE and the shorter, more frequent breaks are best as opposed to longer, more infrequent ones.

The Regulations require that you provide eyesight tests for users if the user requests it. You may need to tell employees that they are entitled to have it done if they are not already aware. An eyesight test would be done preferably before a person becomes a user, and further more at regular intervals. You may also need to provide glasses for users where needed, if their normal glasses won't do and/or they need them to carry out their job as a DSE user. You cannot force someone to have a test. Do ask for guidance on the suggested frequency of repeat tests from the optician or qualified medical practitioner that you use.

Users must be given training on the risks associated with DSE and the recognition of problems, such as how to use DSE properly, what to do about glare on the screen, fatigue, adjusting the equipment, taking regular breaks, identifying health problems and the layout of the workstation. There should also be information on the control measures in place, and what extra action may be needed (such as the provision of foot and wrist rests, for example).

It is important to pay attention to laptop users as they may be using their DSE in unusual, unsuitable and cramped conditions, perhaps in cars, hotels and trains etc. They should be given the necessary information to enable them to set up their laptop correctly and to work comfortably.

Personal Protective Equipment Regulations 1992 (as amended)

Personal Protective Equipment, or PPE as it is most often referred to, encompasses all equipment intended to be worn (or used) to protect a person from one or more risks to his or her health and safety at work. This includes items such as safety glasses, safety footwear, boiler suits, high-visibility clothing, gloves, gauntlets and face shields, and also clothing that is worn to protect against substances, fumes and even the weather. The Regulations do not apply to clothing that is ordinarily worn by the worker as everyday wear - it relates specifically to that which is used to protect against given risks.

Every employer must ensure that they have provided their employees with suitable PPE where they may be exposed to a risk to their health or safety (as determined in a risk assessment). Every self-employed person must also make sure that they provide themselves with suitable PPE if they are likely to be exposed to risks to their own health and safety. This is unless the risk has been controlled to an acceptable level by another means –in essence, companies should really be looking for PPE to be the last resort in the hierarchy of controls. For example, it might be better to enclose the process or control it 'at source' rather than make each individual in the vicinity wear a facemask. In that way, it protects everyone in the area and not just the individual. The relevant risk assessment should make reference to the qualities needed from the PPE to make sure it is efficient against the risks, and it should also identify any risks that the equipment itself may contribute to. This should ensure that the best form of PPE is then identified and used. Make sure that in purchasing PPE, you look at the needs and demands of the work activity that it is for, the health of the employee it is for as some PPE may not be suitable for all people, the length of time that the employee will be wearing it, the weather conditions it will be required to undergo, the fact as to whether it can it be adjusted, and whether it is the most appropriate equipment for the job.

The problem with PPE is that, whilst it can be a very useful protective measure, it relies on the individual to wear it, and to use it or put it on properly. Some people may not be able to wear it or may find it difficult to use it, for example if they have certain health problems, or even wear glasses, in the case of safety spectacles or ear defenders. People are different sizes and shapes, and one size may not fit all – the process of acquiring PPE should involve employees (after all they are the ones who will be wearing it) and as far as possible, you should offer alternatives that will give equal protection. In selecting PPE as a control measure, ask yourself the following questions: Can you ever really rely on someone to wear their PPE 100% of the time? Are you always there to supervise it? Have employees been told WHY they are wearing it, and the risks to their health and safety if they don't? Do they know how to look after it, and when to wear it?

Employees have a duty to wear the equipment that you have provided, and to look after it. They must also report any defects or problems with the PPE, and if they have lost an item so that it can be replaced. Obviously, there is never going to be a never-ending supply of PPE in a company but try not to make it a chore to report losses − it is more important that the employee is protected from the risks as opposed to being disciplined for misplacing a pair of safety goggles, for example.

It can also be hard to determine the level of protection that certain items of PPE offers − but there are better odds on it working (and being used) if it fits properly, if it is as comfortable as possible (several different options may need to be tried), is used properly (so employees have been trained in the correct use of their PPE) and it is maintained well and kept clean. Do check that it is CE marked and complies with the Personal Protective Equipment Regulations 2002 as this means the equipment will have had to have passed vigorous safety tests to conform to certain standards. Employers and the self-employed must make sure that all equipment is maintained, is in good working order and is replaced as necessary. It can be a good idea to get all employees to look at their own PPE − although they will need training and instruction in how to do this - and to not make it hard for them to obtain new ones. A schedule or tick-list would perhaps ensure that nothing is missed during the inspection. Any maintenance can include testing the equipment to ensure that it performs to the manufacturer's standard − the manufacturer may be able to help you with that, depending on what the PPE is. Look at the maintenance instructions that come with the equipment to see how it needs to be maintained. Look also at how the PPE should be stored- can you give it a designated place, where it is protected from dirt and contamination?

As we've stated before, the PPE must be fitted correctly, and must do the job effectively for which it is being used. It mustn't create a new hazard by limiting the vision, hearing or awareness of the wearer. Sometimes more than one type of PPE has to be worn at the same time, for example safety glasses and a hard hat, and you must make sure that these are compatible and as comfortable as possible, i.e. the protection given by one type of PPE is not compromised by the other.

Look at whether or not the equipment is for an individual's personal use, or whether it is to be shared. Cross-contamination can be a factor, especially if it involves the equipment touching the ears, ears or nose. It may be deemed ok for visors or safety goggles to be shared with limited use, but it may probably not be suitable for face-fit masks to be shared, for obvious reasons that everyone's face is not the same shape, and the mask will have come into close prolonged contact with another person. Items such as eye protection should ideally be issued on a personal basis.

Some may, of course, have prescription lenses fitted which prevents them from being shared by others. There is also the question of disposable PPE, and the types of activity that it is suitable for – make sure you check before you buy. Do ensure that users know the time that disposable PPE can be used for, and where it should be disposed of after use. Make employees aware of the limitations of use, and make it clear that disposable means disposable after use!

For religious or other personal reasons, people sometimes declare that they do not want to wear the PPE that they have been given. I once went to a factory where many of the women insisted on wearing flip flops whilst carrying heavy boxes to and fro in the warehouse. If the risk assessment determines that that is unacceptable, then you are within your rights to request that they wear the safety equipment provided for their use, although it is best if you can show that you have tried and subsequently failed to come up with an equally effective measure in its place. Until recently, turban wearing sikhs were only exempt from the need to wear head protection on construction sites-now, however, this exemption also applies to other workplaces, bar some high hazard operations such as certain roles within the emergency services. It is therefore important in this situation to ensure that the risk of head injuries are controlled as much as possible by other means. A note on beards - they can affect the fit of any facemasks and breathing apparatus and a proper face-fit on a regular basis is needed with these pieces of equipment to make sure that they are still working effectively. As with other arguments about personal preferences, although you need to take every measure to adapt and find a suitable alternative, you are still within your rights to request that, for the purposes of the job and to protect that worker in the carrying out of that job, that cooperation be given by the employee to ensure his own safety. These Regulations do not apply where there are more specific Regulations covering PPE, such as noise, as noise issues and PPE are covered under the Control of Noise at Work Regulations.

As previously covered in the section on the Health and Safety at Work Act, you can't charge for PPE provided for use at work.

The follow-on regulations to these are known as the Personal Protective Equipment Regulations 2002 which deal with the design, manufacture and supply of PPE, including CE marking.

Provision and Use of Work Equipment Regulations 1998 (PUWER)

These Regulations exist to make sure that any equipment that is provided for and used for work does not affect the health and safety of employees and others. The Regulations, as usual, apply to employers and the self-employed. They do encompass lifting equipment too, although this is mostly covered by The Lifting Operations and Lifting Equipment Regulations (see relevant section.)

The key point is that risk assessment, as ever, will identify the hazards, risks and control measures needed in terms of the work equipment used at work. The definition of work equipment is very broad – it covers everything and anything from hammers, lifting equipment, laminators, tools, machinery, ladders, and power presses to name but a few items. It doesn't however apply to work equipment provided for use by the *public* in a work activity, though, such as escalators.

The Regulations require that you ensure all work equipment is suitable for the purpose in which it is used or is to be used, taking into account the working conditions, the environment in which it is used and the findings of the risk assessment which will have identified any hazards. Will the work equipment pose any additional hazards, because of the way in which it is used, or the area in which it is used? It should always be used in accordance with the manufacturer's guidelines.

It is important to look at the ergonomical aspects and how suitable it is for persons to use – i.e. is the work equipment heavy? Do you need to exert too much pressure? Should it in fact be substituted for a safer option? I.e. Should you perhaps be using scissors rather than knives for a particular activity? In terms of equipment that is in a fixed position, is the location suitable? Some equipment such as electrical items may not be suitable in bad weather conditions, for example, and therefore should not be kept outside. What about the space around the equipment – is there enough room to move around? Is it in fact blocking a fire exit? What about ventilation, if the equipment gives off fumes or heat?

All work equipment must be maintained in good working order and it should be repaired as necessary in the event of breakdown or fault, and it should be inspected as appropriate. It is often ideal to have resources in place for both planned and breakdown maintenance. Planned maintenance can often lead to a reduction in un-planned breakdowns, as many of the components which could wear out will have already been replaced, such as blades which could go blunt, lubrication, and linings which need replacing. Any maintenance logs should be filled in each time to keep a record of what has happened, and to show proof that the maintenance has taken place. Keep logs and records in a safe place which is easy to find, and ensure that plant is easily identifiable from other equipment, especially if you have more than one of the same items. If work equipment is taken out of service for any reason, make sure that there is a system of necessary checks to certify that it is safe to be put back into production, and that authorisation has been given for it to now be used. No work equipment should be used if it has a defect which could harm someone. Find out from the manufacturer or supplier how often maintenance is needed, and draw up a schedule detailing how this will be done, and when. Ensure that work equipment meets all legislative requirements when you first buy it or hire it in, including those defined under the Supply of Machinery (Safety) Regulations 1992 (as amended) and that, in most cases, it is CE marked. (A CE marking indicates that the manufacturer is claiming all relevant legal requirements have been met – this is just their word though - you still have a duty to ensure that the equipment in your undertaking is safe.) If the equipment is second-hand, some modifications may have been made to it and/or it may have been manufactured before these requirements came in.

If you are hiring any work equipment, ask the hire company before you take it on as to who will undertake responsibility for the maintenance requirements and how this will be accomplished. (It should only be undertaken by those competent to do so, with adequate knowledge and experience.)

Work equipment is required to undergo an inspection before it is first used and at suitable intervals which, depending on the type of equipment it is, may be done by each user, or daily/weekly or at each change of shift. The manufacturer may give suggested timescales for this. Some plant will also be covered by thorough examination and testing requirements via other legislation such as Local Exhaust Ventilation (LEV) which is under the Control of Substances Hazardous to Health Regulations.

One of most important part of these Regulations is Regulation 11 which concerns the dangerous parts of machinery.

Under Regulation 11 you are required to:

Prevent access to any dangerous parts of machinery or stop the movement of any machinery before a person could become hurt by it. Depending on the work equipment this is ideally complied with firstly by using fixed guards, then other types of guards if this is not possible, and then by using things like push-sticks to keep human parts away from dangerous areas. Guards should not be easily by-passed nor should they restrict a person's view of the equipment or its parts if it is necessary to be able to see them to complete the work activity safely. As ever, the risk assessment will determine the type of guard or preventative measure needed.

There should also be controls (buttons, switches or otherwise) where necessary to enable the machine to be brought to a complete stop, and emergency stops placed at appropriate, visible places which do not put anyone in danger when trying to access them. Remember that emergency stops are not a substitute for guards. There should also be a way to isolate the machine or plant from its sources of energy.

PUWER also applies to mobile work equipment which is equipment that carries out work when travelling, or equipment that can travel to different places to do work. It is important that no one is carried by it unless it is suitable for carrying persons (i.e. it is designed for that) and has features to protect the safety of those being carried. Equipment must also be prevented from rolling over and where applicable should have a ROPS (roll over protective structure) to prevent a person being crushed in event of a roll over. A restraining system such as a seatbelt should also be used too if applicable.

Do make it clear who can use the equipment and who can perform the necessary repairs and maintenance. Any one using the equipment must have had suitable and adequate training and instructions for doing so, along with the usual health and safety training on activity, plant and associated risks. To keep unauthorised staff from using the equipment you can supervise its usage, and even put notices on the machine if necessary to make it clear who can use it.

Notes on Guarding

A particular bugbear of mine used to be defeated interlocks. This cannot be stressed enough - DO NOT ALLOW ANYONE TO DEFEAT ANY INTERLOCK - ensure you have a maintenance schedule which involves checking this. Interlocks are there to prevent accidents and save lives, and defeating them, well, just defeats the object. Interlocks can differ in their make up but basically exist as male and female counterparts that are linked together which, when pulled apart, cause a shutdown in the system and make the machine cut out as it receives no power. They are normally found on CNC machines, larger tools which are enclosed and some doors (such as enclosures around the back of guillotines.) The defeated interlocks that I have seen were usually over-ridden due to the desire to cut corners, so that a person could work on a machine whilst it was running, or after maintenance sessions whereby the interlock has not been put back properly. Fatalities and major injuries have occurred where an interlock has been defeated. Enough said.

Whilst we are on the subject of CNC machines and other such large plant, I must mention vision panels. Most enclosed tools have a vision panel so that the worker can see what is happening inside the machine whilst the work is taking place. This panel is made of a strong material which is also designed to reduce the impact of any flying materials which may be ejected by the machine. Over time, swarf, cutting fluid and general wear and tear reduces the effectiveness and strength of this panel and it may become cracked or damaged.

DO NOT REPLACE THIS PANEL WITH ANYTHING OTHER THAN THE MATERIAL DEFINED BY THE MANUFACTURER.

Any old bit of plastic or polycarbonate will not do, and will certainly not contain the impact of any ejected materials such as chuck parts. Fatalities have occurred when vision panels have been left damaged or have been replaced by inferior materials.

Getting caught in machinery

Thankfully it doesn't seem to happen so often now, but sometimes ties and ponytails can get caught in machinery, especially rollers. Watch out for long, wide sleeves of overalls too, and dangly jewellery. If this is a possibility due to the nature of the machinery that you have, then you might want to think about implementing a 'safe machining' policy which details what can and cannot be worn when operating certain plant.

Qu. I have a really old machine which pre-dates 1992 and has no guards on it to prevent access to its dangerous part. What should I do?
The machine mustn't be used until it is safe for use. Undertake a risk assessment to identify the hazards and what could go wrong. Try to contact the manufacturer to see if they can provide you with a suitable guard for that machine. If not, you may need to have a guard retro-fitted by someone else, or if you are in that line of business then you may be able to manufacture one yourself. The important thing to remember is that the machine must be safe to use, the access to the dangerous part must be prevented (preferably by a fixed guard) and the new guarding must not present any new hazards. If you are not sure what to do, you can ask your local HSE Inspector who will be able to give you some guidance on the equipment in question.

Manual Handling Operations Regulations 1992 (as amended)

Manual Handling in the workplace is now a serious problem - musculoskeletal disorders (MSDs) account for around half of all work-related ill health. Commonly these are known as strains and sprains, back pain or back, neck and shoulder injuries. These can happen in any workplace if the proper controls are not in place – it therefore doesn't matter if you are just carrying boxes around the office, pushing loads in a warehouse or even digging up roads – the same types of injury can happen whenever and to anyone. As usual, the Regulations declare that the self-employed have responsibility for their own safety when it comes to manual handling.

Contrary to popular belief, manual handling does not constitute just lifting or carrying something – it also applies to actions such as pulling, pushing, twisting and lowering of a load. A load in terms of these Regulations is a 'moveable object' (and this includes people, such as patients in hospitals etc). There are no set maximum weight limits as to what we should all be carrying, because as individuals we are all different in terms of our capabilities, weight, size and reach, although 25 kg has been used as a guideline in the Regulations as to the maximum weight of a load for a man holding a load at hand height. The Regulations require the following three actions:

- Do a manual handling assessment for all types of manual handling operations carried out as part of a work activity (also looking at any individual factors of the people involved) If you are stuck for where to start, look at the company's accident book – this will help to identify common manual handling injuries that have happened and the work activities in which they occurred. Identify the hazards involved, and the control measures in place and those hazards that are not adequately controlled.
- Reduce the risk of injury
- Give information and training to all those undertaking manual handling work (including the weight of the load that they will be working with, and the hazards that are involved and how to prevent injury to themselves and others.)

The assessment should pay attention to the following, as a minimum:

-The reach distances required during the work activity

-Are there any cramped areas which may require stooping or crouching?

- Are there any loads which are an awkward size or shape?

- Are there any repetitive actions required?

- Are there any heavy loads involved, or is straining likely to take place?

- Is manual handling a frequent and large part of the activity?

- Are twisting actions involved?

- Are persons required to lift loads above their head?

- Are loads to be carried far? Is there any need to change grip?

- Is the load likely to move in transit? Will it become unstable?

- Are there any other risk factors involved such as working at height or slippery floors?

- Have sufficient breaks been allocated?

- If there is more than one person involved, do they know that each person is unlikely to carry the same weight and therefore the weight of the load is not 'halved'

- The nature of the floor – are slopes and steps involved in the route?

- Is the lighting sufficient? Can people see ahead of them?

- Is the temperature adequate? How much time will be spent outside?

Remember to look at

- The task

- The load

- The location

- The people involved

Talk to employees and work through each activity to find out where manual handling operations take place. Identify any shortcuts which are taken for whatever reason, any unsuitable equipment used to complete the activity, and any potential hazards that they have come across such as uneven floors.

Ideally, you should try to fit the job to the person. Is the person undertaking the job suitable to do the task? Do they have the necessary knowledge and training to do it?

Look at the psychosocial factors which can affect the environment in which the employee is operating- is the work stressful? Are there strict deadlines to meet and therefore pressure to get the job done? Pressure can make people work more quickly or haphazardly which can give rise to accidents and injury.

A lot of thought should be put into selecting the best control measures that should be put in place, with a preference given to avoiding the need for manual handling, perhaps by mechanising the process or by using mechanical aids such as trolleys or hoists.

The risk assessment should also make reference to the remedial action needed, along with who will be doing it and by when, and the date when the assessment needs to be reviewed, unless changes occur in the meantime.

Finally, ensure that employees use the control measures put in place.

****

Chapter 4

From Asbestos to Zoonoses: General Health and Safety Topics

Chemicals, Substances and Explosive Atmospheres

Hazardous Substances

Key legislation: The Control of Substances Hazardous to Health Regulations 2002 (amended 2004). Also known as the 'COSHH Regulations.'

Under the above legislation dutyholders are required to do a risk assessment for all substances and chemicals they use at work which may affect the health of employees and others, and to control exposure to them. It includes substances like paints, glues, cleaning products, chemicals, some substances produced in a work process (even as a by-product), biological agents (which are micro-organisms such as bacteria that can cause viruses, allergies and diseases) and certain dusts. Also included are fumes that are produced in a work process – such as those found in welding, or when mixing certain chemicals together. Many hazardous substances can have nasty side effects on the body and human health – some are known or thought to cause cancer, some can cause asthma and breathing difficulties and many effects that substances can have on human life can be long-term or permanent. COSHH assessments do not need to include substances dangerous to safety (such as explosives and flammables) as these come under The Dangerous Substances and Explosive Atmospheres Regulations (DSEAR). Asbestos and lead are also not included in the Regulations as they have their own specific legislation.

To start with, you initially need to identify any hazardous substances that you use, and then look at the information which accompanies the substance. This is called the material safety data sheet (MSDS) and a substance will usually have one of these if it is classified as dangerous to health. To see if a substance has this classification, the label of the container will normally have a hazard sign on it – this is a black and orange symbol depicting an image (think of the well-known skull and cross-bones picture) with a short description underneath it detailing the hazard (*irritant, corrosive,* for example.) Don't forget that we are only looking at substances dangerous to *health* as opposed to safety; those that are flammable or explosive, for example, require a different assessment of risk which can be found in the DSEAR section. Secondly, find out what else you may use or produce, which may not necessarily come with a manufacturer's material safety data sheet. Does anything occur naturally in the workplace, which is potentially hazardous to health?

An MSDS is not a COSHH assessment. It is the information on a substance that the manufacturer is required by law to give you. It tells you the properties of a substance, the first aid actions to take, and the likely effects on the body, amongst other things. It does not tell you what can happen once it leaves the manufacturer and is stored and used by you in your work processes.

The COSHH assessment therefore is an assessment of the risks concerning that substance in your work activities, work conditions and your premises, and therefore your COSHH assessment of a certain substance may well look quite different to that of a another company's.

Often the most difficult part is working out which substances and processes you need to do COSHH assessments for.

Once you have completed them for all hazardous substances, there are many benefits that may arise in the future as a result of having done them – often, and most importantly, there will be less ill-health in the workplace as a result of a better understanding of the potential hazards and effects on the body, you will often see that you are using less materials as your containment measures are put in place and the substances are less likely to fall to waste or escape, and the more robust the controls in place, the less likely it is that time and money will be lost due to system failures and lost working time.

The COSHH Regulations require that, where substances hazardous to health are present, the following points are actioned:

- The risks associated with those substances are assessed, and suitable control measures are put in place to reduce the risks to as low as reasonably practicable. You should ensure that no work takes place involving the substance before the assessment has been done. Either exposure to the substance must be prevented or it must be adequately controlled with systems in place to check that all control measures are working properly. You must also ensure that the control measures put in place are used and not defeated or bypassed by anyone.

- If appropriate, you should provide health surveillance. This is a health assessment by a suitably qualified health practitioner of an individual exposed to a particular substance. You should identify the work activities that require health surveillance to be provided, by looking at the findings of the risk assessment and by seeing if the substance is listed in the COSHH Regulations as requiring health surveillance provision. The Regulations also specify that some other substances not listed may also require health surveillance for individuals if there is an identified disease or condition associated with use of the substance, and there is a likelihood that the disease will occur under those working conditions in which you use it. The Regulations also give guidance on the suitable level of health surveillance that should be given, ranging from visual inspections, questionnaires on health, lung function tests and invasive tests such as blood tests. Work out who will arrange the health surveillance and where the records will be kept. There are also some types of health surveillance which need to be given at strict intervals for certain substances so someone within the company will need to liaise with the health surveillance provider to make sure that these timings are kept to. You must keep all records for all individuals for at least 40 years.

- If appropriate, monitoring should also be undertaken. This is a useful tool to check if your other control measures (such as local exhaust ventilation) are working as it measures the level of substance that individuals are exposed to in the air. Obviously, this will only work for airborne substances. You should identify any jobs that may need monitoring, then work out how it will be done and what measuring device you will use, who will do it and what records need to be kept. Generally, if there is a Workplace Exposure Limit (WEL) indicated on the material safety data sheet then you know that there is an approved maximum level of substance which an employee is allowed to be exposed to in a given time. The information you get should be compared to the maximum exposure limits for each substance which can be found in the HSE publication EH40. This publication also gives the WELs for substances which may not have a manufacturer's material safety data sheet. In essence, monitoring should be considered if there is a possibility that control measures might fail, if it is possible that individuals may exceed the WEL assigned to the substance, or if serious health problems could occur from the substance.

- Information and training must be given to employees on the hazards associated with the substances, and the control measures that you have put in place. Specific health and safety training must also be given to individuals on the substances that they come into contact with, as to how they must use it safely and what precautions they must take, such as the appropriate PPE to be worn.

- There must be procedures in place in case of accidents and emergencies. Ensure that you communicate these procedures to employees and others who may need to know and, if appropriate to the degree of risk, the emergency services. Detail what alarms mean and any other warning signals, and the procedures of evacuating the premises in the event of an emergency.

In doing a risk assessment for hazardous substances, look at the following as a minimum guideline for what should be considered in the assessment (a lot of the information that you will need will be displayed on the material safety data sheet):

- What is the substance? What are its ingredients and hazardous properties? Who is the manufacturer or supplier that you got it from? Does the substance have any other names? What form does the substance take, i.e. is it a liquid, solid or gas?
- What processes and work activities is the substance used in?
- What quantities is it used in (and stored in, if different)? How often is it used?
- Who could be affected by it, and how? Are there any additional hazards for certain types of groups such as new and expectant mothers? Who else is in the vicinity that could be harmed, who is not an employee?
- What are the possible routes of exposure – could it enter the body via the skin, via inhalation, by injection or by ingestion?
- What control measures are in place, and what more needs to be done? Look at the hierarchy of controls, which includes, in order of priority, getting rid of the substance (substituting it where

appropriate for something else), segregating it so that there is an enclosed process, controlling it at source (i.e. with ventilation/extraction), giving information and training on the hazards, supervising workers and using PPE. Whatever control measures are used, it is imperative that good systems of work are in place, with safe working procedures that are known by employees. It is also important that any Workplace Exposure Limits are not exceeded, and that equipment involved is examined and tested as appropriate and maintained in a good working condition.

- If applicable, what is the Workplace Exposure Limit (WEL)? How do you know that it is not being exceeded?
- What are the 'risk phrases' listed on the MSDS – they indicate the classifications of health effects
- Is Health Surveillance needed? What type, and how often?
- Is monitoring needed? How will this have been done, and how often should it take place?
- A clearly-defined procedure to deal with accidents and emergencies should be in place if the assessment deems it necessary.
- Are any biological hazards (micro-organisms that can cause infection or allergies amongst other things) involved? Biological hazards are classified into zones and you will need to ensure that you comply with the extended requirements for biological agents if they are present in the workplace.
- What are the harmful effects of this and any other substance when used together?
- Is there a possible substitution or alternative that could be used instead?
- What are the first aid measures to be taken?
- Where is the substance stored or present – are there any special precautions to adopt?
- Are there any further safety precautions for using the substance?

- Is any further investigation needed? Are all the hazards deemed to have been identified? Are any further assessments required, for example under other legislation?

The person or persons doing the assessment should be competent and should know and understand the requirements of the COSHH Regulations. They should know what to look for, and should be able to make decisions about control measures and actions to be taken, or at least have access to someone who does. The assessor should also consult with employees, especially on usage and the things that can go wrong, as these are the people who will probably know this best. The assessments should be recorded if there are five or more employees (you can do this however you want, be it in hard or soft copy, but they should be readily available and, like the material safety data sheets, they should be kept near to where the substance is used for employees to refer to). They should be reviewed periodically, and if changes occur, or the results of health surveillance or monitoring show that control measures are not working properly.

Occupational Health Provision

If you use an Occupational Health Provider to undertake your health surveillance or screening, do look at the Service Level Agreement if you are given one, to be sure that you know exactly what you are getting for your money, that persons within your company know what to do with the information they are given from the provider about individuals, and that they know how to refer individuals who have work-related ill-health.

If employees have a work-related condition then you must ensure that these conditions are managed effectively and the employee is not exposed to any further risk which may exacerbate their condition. Depending on the severity of the condition, they may not be able to work with that substance and therefore may need to be given alternative work.

Finally, there is a free website provided by the HSE called COSHH Essentials which takes you through the process of completing a substance risk assessment. It can be found at www.coshh-essentials.org.uk

REACH - Registration, Evaluation, Authorisation and Restriction of Chemicals

REACH is an EU directive which came into force on the 1st June 2007.

It requires that companies which manufacture or import more than **one tonne** of a chemical substance per year into the EU need to register it on a database provided by the EU Chemicals Agency. It has three main reasons: to reduce the amount of animal testing (over time); to try to phase out the most hazardous substances to health and the environment by encouraging the switch to safer substances which require less documentation; and to ensure that as much as possible is known about the chemicals in our environment. REACH applies to individual chemicals, on their own or in a mixture, as long as they are present in a weight of one tonne or more per annum.

Companies who manufacture or import substances over 10 tonnes per annum will also need to prepare a Chemical Safety Report. This will be a detailed description of the hazards and control measures needed when using the substance and the information contained within this, and the initial registration documents, will be disseminated down to the users of the substance to ensure they have as much information on it as possible.

Duties for registering substances will be shared as the EU Chemicals Agency does not want multiple registrations and will therefore encourage and facilitate communication between companies to enable them to get together and submit information between them – they will then also be able to share the fees as it will cost companies to register.

REACH will also introduce a new authorisation scheme for 'high risk' substances, such as carcinogens (substances known to be cancer-causing) and those that may induce birth defects, amongst other things. Controls will be tightened over the years, and the burden of proof will be on industry to demonstrate that the risk from the use of a chemical can be adequately controlled, and to recommend appropriate control measures.

For all new substances, registration starts on 1st June 2008, and from then on substances need to be registered before they can be put onto the market. There will apparently be advantages for SMEs, as they realise that this group may be more likely to incur problems from not having the resources as a lot of work may be required in the beginning and therefore SMEs are likely to benefit from lower registration fees. Also, as they are most likely to be registering small amounts, SMEs will most likely fall into the category of not having to register for 11 years after REACH comes into force as the need to register different tonnages is staggered over more than a decade.

Companies that use the substances and do not manufacture or import them will be deemed 'downstream users' and will have duties of their own – which include the requirement to tell the manufacturer/supplier of *their own* uses of the chemical to ensure that their use of it is covered in the registration documentation submitted. Another requirement is to update any relevant COSHH assessment information as more data and information on the hazards and control measures is likely to be provided from the manufacturer.

REACH will be controlled by the European Chemicals Agency (EChA) who intend to develop free IT tools that can be downloaded to help with the burden of registering many substances, and to facilitate information gathering and the passing on and sharing of information up and down the supply chain.

Dangerous Substances and Explosive Atmospheres
Key legislation: The Dangerous Substances and Explosive Atmospheres Regulations 2002. Commonly known as 'DSEAR'.

Whereas the COSHH Regulations concern substances hazardous to health, DSEAR relates to all dangerous substances which are hazardous to safety. Dangerous substances are defined as those that could, in the right conditions, cause a fire or an explosion. Examples of these substances include flammable solvents, some paints and glues, some dusts that can create 'dust clouds' that can then be ignited, and some petroleum products. (In 2014 the Petroleum (Consolidation) Regulations came into force that concerns where petrol is stored and dispensed in work premises and other places. Petrol filling stations now need to hold a Petroleum Storage Certificate.) Even small amounts of these various explosive substances mentioned
could create a hazard, and therefore DSEAR may apply to more companies than you might think. Indeed, even small factories and offices may have cupboards with paints and glues in them, and beauty salons have nail varnishes and solvents. Fires and explosions can obviously cause severe damage and human injury, even death. Multiple persons could be affected, even if they are not immediately hurt by the impact, as fire can reduce the oxygen in the vicinity and people can suffer from asphyxiation.

Under these Regulations there are certain requirements for dutyholders to fulfil:

- Ensure that the risks for all dangerous substances (either used, or produced) that are hazardous to safety are assessed, and that control measures are put in place to either remove the risks or reduce those risks to as low a level as possible. You should therefore identify all substances which come under this category, and ensure that all associated hazards have been identified. Look at the MSDS to see what information on fire and explosion that this contains, to aid your assessment. You will also need to identify any possible ignition sources that could start a fire or explosion. For a fire to start there needs to be a fuel, an ignition source and oxygen, and in this case the fuel would be the dangerous substance. It is obviously impossible to prevent oxygen occurring in an area, apart from a vacuum, and therefore it is the ignition source which must be removed as a possibility of occurring. An explosive atmosphere could build up over time, until there is enough of a substance in the air. Remember that some fine dusts, gases and vapours cannot be seen or smelt and therefore could go undetected.

- Put measures in place to reduce the damage and effects that a fire or explosion would cause. How could you stop a fire from escaping to another area, for example? Also, could an explosion lead to a fire starting, and therefore both could happen?

- Give information and training to employees on the hazards associated with the substances and the findings of your risk assessment, so that they understand the risks and how they must use the substances and any control measures you have put in safely. Employees and anyone else who is likely to be affected by a fire or explosion from dangerous substances should be informed of the plans and procedures in place should an adverse event occur, and what is required of them either to help mitigate the effects (only trained and specialist personnel), and/or how and where they should evacuate to. If you have dangerous substances on the premises on a large scale, and there is potential for a fire or explosion because of the nature of the work, the emergency services may also like to know in advanced the likelihood of a fire or an explosion occurring on your premises, and any information on the substances that you hold.

- If dangerous substances are in an area, and there is potentially a chance that these substances could be ignited by whatever reason, then that area should be classified and labelled as an 'explosive atmosphere' area (known as ATEX) which is visibly seen before entering the area and further divided into different zones relating to the likelihood of fire and explosion occurring. See the Regulations for details.

- There is no need to take things further than implementing control measures if there is deemed to be no or low risks from the substances.

When risk assessing dangerous substances, you might like to look at the following as a minimum:

- What are the hazardous properties of the substance? Why is it dangerous to people's safety, and under what conditions could it give rise to a fire and/or an explosion? How is it categorised on the manufacturer's material safety data sheet MSDS) – is it classed as flammable to any degree, oxidising or explosive?
- Who might be hurt, and how?
- In what work activities are dangerous substances found? How is the substance used and stored, and in what quantities? Is the way in which it is used or stored a factor in its classification as a dangerous substance? i.e. is it not normally a dangerous liquid until it is heated to above a certain temperature?
- Is there any dust which occurs naturally or otherwise, such as from flour which can form a cloud that can be easily ignited?
- What are the temperatures and pressures found in areas where dangerous substances are present – could this affect them?
- If multiple substances are used in an area, could there be a knock-on effect? What about substances reacting with one another?
- What is the potential for accidental release of a substance? What could happen?
- Look at the necessary control measures and precautions for the safe handling, transport, delivery and storage of substances.
- What is the likelihood of ignition source, and explosive atmosphere occurring at the same time? What would the ignition

67

source be likely to be? Consider heated surfaces, flames, welding, static from humans or plant, and lighting that is not compatible with these sorts of environments.

- What are the anticipated effects of a fire and or explosion? What is the worst event that could happen and why?

- What are the safety measures that are in place, and what else is needed? Firstly, the most important thing is to prevent an explosive atmosphere happening in the first place. It may be necessary to reduce the amounts of the substance that you use and store on site. Then there should be containment systems to enclose substances and prevent them from escaping or building up in the air. Local exhaust extraction may also be needed to remove substances from the work area. Suitable PPE should be available generally, and in the event of an emergency. Training and instruction should be given on the hazards and risks concerning dangerous substances, and the use of control measures. It is also important, if an explosive atmosphere could form, that there isn't anything in the area which is not resistant to explosion, or fire.

- What are the fire safety practices, which are used within the company?

- Do areas need to be zoned?

- Any other information which may need to be recorded.

As with all risk assessments, they should be recorded if you have five or more employees and they should be reviewed periodically, or as any significant changes occur.

Major Accidents and Hazards

Key Legislation: Control of Major Accident Hazards Regulations 1999, amended 2005. Commonly known as the 'COMAH Regulations.' For some companies which hold or produce large amounts of dangerous substances on site there are another set of Regulations which can apply under certain conditions. A list of these substances can be found in the COMAH Regulations which contains some named substances, and some categories of substances (such as flammables). The COMAH Regulations only apply if certain quantities in tonnage are exceeded. The principles of COMAH are to prevent, and mitigate against, major incidents involving fires and explosions and emissions of chemicals into the air which could have a serious impact on the environment.

There are lower and higher thresholds for the substances, which depends on the amount of the substance in tonnes. For substances listed which are in quantities belonging to the lower threshold this involves notifying the local HSE office and preparing a Major Accident Prevention Policy (MAPP) and reducing the risk to as low as possible. If the higher threshold applies, then the same action must be taken along with preparing items such as emergency plans, safety reports and making the information about the hazards available to the public to see.

COMAH sites will be those that have the threshold quantities of the substances *in one location owned by the same person or body*, be it in storage or use, or as produced in a chemical process. So, if you have two factories, with one being in London and the other in Reading, you do not have to add together the quantities that you hold on both sites and therefore each site will be classed as an individual location. Look at schedule 1 of the Regulations to see if you fall into any of the categories. Some chemicals can react together to produce new chemicals, and these may well be listed.

Lifting Equipment

Key Legislation: The Lifting Operations and Lifting Equipment Regulations 1998 (commonly known as 'LOLER'.)

Over the years there have been some catastrophic failures of lifting equipment, involving plant such as cranes and mobile elevated work platforms that have lead to severe injuries and fatalities of the workers involved. Often, it has been found that these pieces of plant have not been maintained properly, and parts have failed which have lead to accidents that could have been prevented had the necessary maintenance been done.

Often accidents occur because unsuitable equipment is being used to lift loads, such as when items have been carried on the forks of forklift trucks which were too heavy and the balance has been over-reached and the forklift has tipped over. It is important that the correct equipment is used for the job, especially when the load is a person.

Lifting equipment is classed as something that is used at work for lifting or lowering loads, and any attachments that are used to help with this such as slings and ropes, etc. The load is the item or artefact that is being carried or moved, and therefore can also be a person or persons. The LOLER Regulations only apply to work activities. Lifting equipment encompasses plant such as cranes, hoists and tail lifts on vehicles, and accessories such as ropes and slings, yet it also applies to lifts used in offices and factories for example. It doesn't include escalators though as these are covered under different legislation. The ropes used for climbing can also be covered by LOLER in some work activities. Check the Regulations if you want to find out if they apply to you and the equipment you are using, although generally if equipment is used for lifting loads and if its primary function is to lift or lower loads then it will probably have to meet the requirements of LOLER. Also be aware that the work equipment and associated requirements may also be included in The Provision and Use of Work Equipment Regulations (PUWER). The Regulations apply, as usual, to the self-employed for equipment that you use at work and any persons who may have a degree of control over lifting equipment

The Regulations state that all lifting activities must be properly planned and carried out by trained and competent employees - especially if more than one piece of lifting equipment is being used at the same time or in the same area as this increases the likelihood of accidents occurring. If employers allow employees to use their own equipment at work, then LOLER will still apply to the equipment. You will still need to ensure that the correct tests, examinations and inspections take place and it is sometimes best not to let this happen as a rule, as rogue items may appear that haven't had the necessary test and examination.

If you hire equipment - such as a crane – from a hire company then they have a duty to provide you with the last examination report and other details regarding the safety of the equipment. On installation, the new user (i.e. the dutyholder) needs to ensure that the equipment has been thoroughly examined by a competent person before it is used. The Regulations state the frequencies at which examinations must take place. On a longer-term hire, it may be best to agree with the hire company beforehand who will do these examinations and testing and any necessary maintenance, and how and when this will take place...

Often the best way for persons to access heights whilst alleviating the need for climbing is to use a Mobile Elevating Work Platform (MEWP) which is an enclosed 'box' attached to a vehicle which is extended to the desired height and is either controlled from below or from within the box. Only that which is suitable and meant for persons should be used to lift or carry people – and that does not mean balancing on the forks of a forklift truck. Do not ever allow this to happen!

A risk assessment for work activities involving lifting equipment would cover, as a minimum:
- The hazards involved, and the control measures in place and those that are needed to further reduce the risk. Think about failures with the lifting equipment, and also if something was to happen to the load. What if the load was to slip, or fall? Consider the weights involved, and how safe working loads will not be exceeded.
- Who could be harmed, and how? Think about the person controlling the lift, anyone that may be near the load and others in the vicinity.

- How will each work activity be done safely, as well as taking into account the cost and time and other resources in buying or hiring suitable equipment? How will the correct equipment be selected - what is the criteria?

- Think about the weather factors, the material that the load is made from, visibility to see where loads are going, slips and trips, and the location of the work area.

- Especially when lifting equipment is involved in lifting persons above ground level, think about access and egress, edge protection, falls from height, objects falling, stability of the ground, overturning of lighting equipment, stopping people falling out or being crushed or trapped or injured (such as having cab doors open inwards when at height), whether harnesses are also needed when working at height, communication with persons on the ground, and how you can stop lifting equipment moving when at height.

Making equipment safe

Thorough Examinations and regular inspections will be needed for lifting equipment to make sure that it is safe for use. Most lifting equipment will need to be under a written scheme of examination (with records kept) by a competent person with the necessary training and experience. It is also necessary to check for faults and defects before first use. Plant must be inspected after it is first installed, and again after it has been re-installed or moved elsewhere. Lifting equipment used to lift people and lifting accessories should be thoroughly examined at least every six months. Other lifting equipment needs to be thoroughly examined at least once every twelve months.

During the thorough examination a report will be produced by the competent person which will detail if the plant is safe for use and if there is a cause of serious imminent danger then they will also be required to inform the enforcing authority of that defect.

You should keep all thorough examination reports throughout the life and use of the equipment until it is no longer in use, and if you pass it on to someone else then you should give them these reports. The reports given for lifting accessories should be kept for at least two years – these items tend to be changed quite often as they become worn out and need replacing, hence the shorter time period for keeping these types of reports.

Occupational Health

As detailed in the section under Chemicals, some substances and materials can have a detrimental effect on health. Employees who use some of the many substances which can cause certain identifiable diseases and conditions may need to under a scheme of health surveillance by a qualified and experienced occupational health practitioner. Details of local providers can be found in the phone book.

Below are some of the major issues in the realm of occupational ill health which can affect persons:

Dermatitis

Dermatitis is a skin condition which causes inflammation, flaking, itching and drying of the skin which comes about primarily from contact with hazardous substances and materials. In some cases it can cause 'chemical burns'. Dermatitis is usually found on the hands as these are the parts of the body that comes into contact with the substance although it can spread to other parts of the body. Dermatitis can occur from repeated contact with substances such as chemicals, solvents, cleaning products, cooling fluid, and even water with prolonged use.

There are two main types of dermatitis – irritant contact dermatitis and allergic contact dermatitis. Irritant contact dermatitis occurs when the

73

substance causes a reaction with the skin at the site where the contact happened. Allergic contact dermatitis is where the person becomes sensitised and allergic to the substance, and therefore the reaction will happen each time and sometimes for a while after the contact. When looking for signs of dermatitis, look for dry skin, blisters, itchy skin, cracks and painful areas, especially between the fingers. It has been known for wet cement to get into builder's boots, for example, which can then go un-noticed for some time, until the skin starts to react, so it is a condition that can often go un-noticed for periods of time, until the changes start to happen in the skin.

What to do:

Do a risk assessment relating to dermatitis and the activities in which it may present itself. Look at the frequency of use and decide who could be affected by dermatitis. Pay attention to the material safety data sheet that comes with the substance and the guidance that it gives. Try to ensure that employees avoid contact with the substance and if possible try to use safer, less hazardous substances. Ensure that appropriate PPE is used (i.e. protective gloves) so that there is no direct handling of the substance. Give employees training on what to look for, and in how to perform hand washing correctly. Have a skin checking system in place with someone who knows what they are looking for. Make sure gloves are provided, and that they are of a suitable type that the substance will not filter through them. Check that employees are not allergic to latex, or the powder in some gloves. Offer the choice of a selection of gloves made out of different materials. Ensure that barrier creams and decent washing facilities are available. Cuts, burns and sometimes the powder in protective gloves can allow for better penetration of the substance into the skin.

A telling sign of a work-related skin condition can be if the symptoms clear up on holidays or weekends. A qualified Occupational Health practitioner should be able to tell what type of skin complaint it is, but the controls will be the same. Once a person has had dermatitis, it is important to make sure that they are not exposed to the substance or a similar substance again, and that suitable controls are in place.

Allergic reactions to Latex

Latex is a natural product which can cause skin irritations in some people, and in others an allergic reaction. Sometimes, but rarely, it can lead to an anaphylactic shock in serious conditions. Latex is used in some types of protective glove, and some medical items, amongst other things.

An adverse reaction to latex may take some time to show up, as a person can become sensitised over time, or it may happen immediately. Eventually, a sensitised person may not even have to be wearing the gloves and can just be in brief contact to experience the symptoms.

Key actions to take:
-Do a risk assessment to control the risks associated with latex gloves. Identify who may be at risk (nurses, car mechanics, beauticians and hairdressers are amongst those groups of people who may often wear latex gloves) Look to see if any problems have been experienced by questioning staff.
- See if latex gloves can be replaced by another type of glove
- If there is a need to wear latex because they are the best suited to job, choose those with lower levels of latex
- Some latex gloves contain powder that can be released into the air and breathed in. This powder can also be a respiratory sensitizer and could give rise to asthma and associated work-related conditions.
- If you decide to use non-latex gloves as a general policy, ensure that no new health risks arise from these new gloves.
- For each work activity, define when protective gloves need to be worn, and when it is not necessary for them to be worn.

- Have a selection of different types of gloves available, with information for employees on the different types of material that they are made from.

- Note that barrier creams may increase the amount of latex absorbed by the body if used together with the gloves.

-Provide health checks for those staff exposed to latex – if health problems have been identified, try to establish that it is latex that is causing them, as it could it be a combination of things that are the problem

- The main thing to note is that it if a person has been sensitised to latex, that they are not allowed to come into contact with it again at work as this could make the problem worse. In this instance you will probably need to do an individual risk assessment for that person to ensure that suitable control measures are in place.

Cancers

One of the biggest causes of skin cancer is the same for people whether they are at work or not – the sun. Employees must be protected from sun damage whilst they are at work, and therefore if they must work in the sun they should be provided with suitable clothing and a place to have a break in the shade. Some companies also provide sun creams to employees if they work constantly in the sun. Skin cancer can also be caused by some substances coming into contact with the skin, such as some types of oil and petroleum. Employees should regularly change their overalls if they work with these types of substances, for example as mechanics may do, and they should not wipe their hands on their overalls, or store oily rags in their pockets. Good hygiene facilities must be provided with hot water, soaps and barrier creams as appropriate.

Radiation can also be harmful in certain doses and attention must be made to the specific regulations covering radiation if this applies.

Some substances are known to be cancer-causing and therefore these must have high levels of control measures in place.

Occupational Asthma

Asthma can be a debilitating condition and in some cases people even have problems with walking a few metres or having a conversation. This can obviously affect a person's ability to do their job, and they may not be able to work with the substances or conditions which have helped to cause it. Substances such as some dusts, and other respiratory sensitizers can cause occupational asthma. It can even happen when not in direct contact with a substance and it may take time for the effects to be seen.

Occupational asthma is often, like dermatitis, an allergic reaction to substances which are airborne.

A risk assessment is required for all work with respiratory sensitizers and high levels of control measures must be in place. Wherever possible, the substance should be substituted with that of a less hazardous nature and/or the process should be enclosed or controlled at source.

COPD (Chronic Obstructive Pulmonary Disease)

COPD is a group term to describe lung disease/lung damage which affects a large part of the population, and is often associated with the effects of smoking. However, combined with the hazards found in certain industries, such as dusts in the painting, decorating and woodwork trades, and certain fumes such as welding fumes, the effects can be exaggerated and the risks increase. Also, people who have never smoked have also been diagnosed with COPD which has been attributed to the fact that they breathed in dusts or fumes etc due to their occupation.

Once the damage has been done to the lungs it cannot be rectified – the only thing to do is to try to prevent it from getting any worse. This can be done by reducing or preferably eliminating exposure to the offending substance, and to stop smoking (including passive smoking). Employers should aim to use low-dust materials and where appropriate introduce wet-working to prevent dusts escaping. Ventilation, extraction and good housekeeping should be used to prevent an accumulation of substances in the workplace. Health surveillance and monitoring may also need to be implemented (see section on COSHH).

Blood-borne viruses (BBVs)

Blood-borne viruses are those such as Hepatitis B and HIV that are carried in the blood but which can be contracted via other bodily fluids such as semen, and also urine and saliva (especially if these contain blood within themselves.) Contact with blood and other fluids is obviously a risk for healthcare professionals and those that work with needles, but it also needs to be thought about for people in the rubbish disposal/recycling industries, construction workers, beauty/tattoo parlour employees and anywhere else whereby people might come into contact with body fluids or receive a needle-stick/sharps injury.

Under COSHH, employers should assess the risk of infection by a blood-borne virus and put the necessary control measures in place to protect the health of employees. If it is deemed to be a risk, then information and training on the hazards should be given, and good hand-washing procedures should be in place. Immunisation may be available for some viruses (such as Hepatitis B). Any cuts or broken skin should be well protected and appropriate PPE should be worn. A good system for the careful storage and disposal of sharps should be used. Care should be taken when doing the risk assessment to see if any other routes of entry into the body are relevant, such as splashes which could reach the eyes, nose and mouth.

Sharps injuries

A new set of Regulations relating to sharps injuries were introduced in 2013 concerning sharps injuries and their prevention within the healthcare sectors. The Health and Safety (Sharps Instruments in Healthcare)

Regulations 2013 require companies within this sector to ensure that they have sufficient risk assessments and associated control measures in place to prevent injuries from sharp instruments such as needles.

Zoonoses

Zoonoses are animal diseases that can also be passed on to humans. The people most likely to be at risk are obviously those that work on farms, in zoos or in veterinary practices, but also those who are visiting these types of workplaces. Some of the associated diseases are not that common, such as anthrax, but others like Weil's Disease (Leptospirosis), which can be contracted via contact with rat's urine, is more common and can be found in watercourses where rats have been. In this case, workers who perhaps work underground, near sewers/water or near areas containing rubbish could potentially be at risk. A COSHH assessment should be undertaken if there is deemed to be any chance that employees could come into contact with zoonoses − if you are not sure you should find out what the various diseases are that employees could encounter. Ask for more information if visits to places where animals are present are to take place. Often those persons most at risk will be those who are pregnant, the very young and the old, and those with poor immune systems or those who have recently been ill or undergone surgery. Keeping up to date with relevant vaccinations, maintaining good personal hygiene such as hand-washing, and wearing appropriate PPE may also be important.

Fire

Key legislation: The Regulatory Reform (Fire Safety) Order 2005

Fire can be deadly, there's no doubt about that. It is therefore paramount that employers take all necessary steps to reduce the risk of fire in their workplace, and the spread of a fire should it happen, and take measures to protect the safety of employees and non- employees who may be on the premises. Buildings, premises and work activities should be designed and

79

planned so as to prevent a fire occurring, and measures should be in place to detect and warn if there is a fire.

The Regulatory Reform (Fire Safety) Order, made in 2005, is a consolidation of nearly all existing fire safety legislation. The main requirements of this piece of legislation are that duty holders have to carry out a fire risk assessment and then act on the findings of that assessment. Fire certificates will no longer be valid but may be useful as a starting point for your risk assessment. Fire rescue services can still inspect your premises, particularly in higher risk premises, but they won't carry out your fire risk assessment for you although they may give you advice and you can certainly ask them for guidance. So, if you are responsible for a business premise or any part of a building used for work, are an employer or self-employed within that premise, are a charity or voluntary organisation, or you are a contractor with a degree of control over a premise then you need to ensure that you have a suitable and sufficient risk assessment in place. The Regulatory Reform (Fire Safety) Order came into force in October 2006 and is the responsibility of the office of the Deputy prime minister who is currently responsible for general fire safety issues.

Northern Ireland and Scotland will have their own laws as this order only applies in England and Wales.

The responsibility for the enforcement of the Order will be with the local fire and rescue authority who may carry out regular inspections of premises. The HSE is mainly concerned with workplace fire safety and process fire precautions that are connected with the work activity carried out, including the safe storage of substances and materials. Process fire precautions are enforced by the HSE and the local authorities under the Health and Safety at Work Act 1974 and more specific legislation such as the Dangerous Substances and Explosive Atmospheres Regulations 2002. (See chemicals section) The main aim of both the fire authorities and the HSE however is to prevent a fire happening, reduce the likelihood of a fire breaking out in the first place, and if it should, to reduce its spread and intensity. Control measures would be those such as adequate ventilation, identifying and preventing sources of ignition, using extraction systems, and storing flammables safely, for example.

Responsibility for complying with the Order rests with the 'responsible person' i.e. in the workplace this may be the employer, and otherwise any other person who may have control of any part of the premises, for example the occupier or owner. In all other premises normally the people in control of the premises will be the 'responsible person'. If there is more than one responsible person then they must work together to ensure that all particulars of the Order have been complied with.

The fire risk assessment must focus on the safety of all persons in the vicinity in case of fire, and particularly those at specific risk such as those with special needs or elderly people. It must also look at substances that are likely to be on the premises and the hazards that may arise from their presence.

The Regulatory Reform (Fire Safety) Order 2005 requires that the following steps are taken:

- An assessment of the risks is undertaken to identify all fire hazards which could occur within your premises and work activities, and the control measures needed, looking at those groups of people who are likely to be most at risk (such as those with special needs, lone workers, children, young persons, the elderly and those with disabilities) and seeing how you would get these people out in the event of a fire. It requires that you look at possible escape routes, evacuation procedures, warning signals (fire alarms) and training requirements for staff. It also requires that you consider those not in your employment – such as members of the public, and other people's employees working in the vicinity. The main aim is to remove and/or reduce the hazards (such as separating flammables from possible ignition sources) and implement good control measures for any remaining risk. It is also a good idea to look at

81

instigating safe smoking policies and to be aware of the designated smoking areas and to disallow any non-designated areas that people tend to smoke in as these could potentially be a fire risk.

- From there, identify the fire precautions needed (such as fire alarms, fire detectors and emergency lighting) and ensure that they are installed, tested and maintained as appropriate. Dangerous substances and materials which may be a fire hazard should ideally, and if possible, be replaced by alternatives which are known to be safer. Fire fighting equipment such as fire extinguishers should be accessible and easy to use, and appropriately labelled. The correct extinguishers should be placed in the vicinity to which they relate, as it is no good having a water extinguisher that might be used on electrical equipment. It is also important to assess how many fire fighting components you will need, which may depend on the size of the building, the amount of people working there at any given time, and the work activities that take place. Automatic fire fighting equipment like sprinklers should be regularly tested to make sure it works correctly.

- Give the necessary information, instruction and training to employees - ensure that the findings of the risk assessment are distributed to all those likely to be affected, and that key personnel (such as fire wardens) know what their responsibilities are in an emergency and practice evacuations should these occur. The findings of any risk assessments concerning dangerous substances should also be disseminated to employees. Training should be given to competent persons who are nominated to undertake the use of any fire-fighting equipment — the fire authority may be able to help with this. The most important thing to remember is that people should only use fire-fighting equipment if they need it to get themselves out, or alternatively if there is no danger to their own safety and it is deemed to be safe for them to do. Buildings can be replaced – people can't.

- The risk assessment should be recorded if you have five or more employees. Review the risk assessment as necessary (at least annually), and in the event of any significant changes such as the numbers of employees, the location of the premises, the processes and the materials and the amounts used, and for any other reason if you suspect that it may no longer be valid.

If you have particularly hazardous work activities on the premises, or dangerous substances which could give rise to a fire, then you could make contact with the fire authority and give them this information so that they have it on file. This might also be a good idea if your premises are in the middle of nowhere or are hard to find. It is a good idea to train people in the basics of how to use a fire extinguisher (see point 3 above, though.) They can be quite dangerous pieces of equipment if not used properly, and, lets face it, no one stops to read the instructions in an emergency, do they? I have heard of people who have dropped heavy extinguishers on their feet when trying to use them, which meant that they then had to be carried out of the building by another person and therefore both rescuer and rescuee were both put at risk.

Fire extinguishers are also pressurized and can explode if dropped or knocked over, which is why they should be hung on the wall on the hooks that they come with, rather than being used to prop open a door or rest your coat on.

Make sure that all fire exits are kept clear from blockages, and that they can be easily accessed and opened in a hurry. You could make it part of someone's job role to check all fire exits on a regular basis. Emergency exit doors must be marked as such and should not be locked if it means that the door will not be able to be open immediately.

Make sure that in the event of a fire, no one goes back into the building until it has been deemed safe to do so, usually by the fire brigade. There should be a system in place so that you know who is on site, such as a logging- in system and a visitor's book. Make sure that people know the importance of making their whereabouts on the premises known, and that visitors are escorted at all times and told the evacuation procedure and where to go in the event of a fire. If you have a large site, there should be some form of communication between buildings to raise the alarm. If you share a premises, or share responsibilities for implementing the Order with anyone else, make sure that you pass on any relevant information and co-operate to ensure that all hazards and control measures needed are identified.

First Aid

Key legislation: The Health and Safety (First Aid) Regulations 1981

In case employees (and possibly others) become ill or injured at work you will need to ensure that you have adequate first aid provision in place, which includes staff with first aid responsibilities, and first aid supplies. This also applies to the self-employed too, who need to have a first aid kit for themselves. You don't necessarily have to provide first aid facilities for others who are not employees, such as members of the public, but it is good practice and it makes sense to, as you would want to be able to deal with any injuries or accidents properly, no matter who it had happened to. You may also have a large number of the public on the premises, for example.

What do I need to provide?

Generally, doing a risk assessment will tell you this. Lower and higher -risk premises require different things in terms of the level of provision, the number of trained staff and the items needed in a first aid kit, the number of kits that should be available and whether or not there should be a specific place designated for first aid to take place (certainly in higher risk workplaces this may be the case.)

Firstly, look at the hazards that you have on site and other risk assessments that have been done for work activities. Is there a likelihood that a major injury could occur, to one or more people? The amount of qualified first aid personnel you may need depends on the severity and likelihood of an accident, and the number of employees that you have. As a guide, you could look at having at least one first aider for every 50 employees in medium to high-risk places. For low risk enterprises, such as offices with only a few employees, you might look at having at least one appointed person. There isn't really a defined set number which says how many you should have- it really does depends on your assessment of what you need and a bit of common sense.

Appointed persons:

These are staff that haven't necessarily had any training (although courses are available and it can help to give them a better understanding of their role if you do put them through a suitable course.) Appointed persons do not give first aid – they are there to help take charge of an emergency situation and follow the first aid procedure- for example, phoning the ambulance and getting hold of one of the company's first aiders.

Appointed persons often take over the responsibility of looking after the first aid kits(s) and ensuring they are kept up-to-date and well stocked with items.

First Aiders:

First aiders have to have completed a recognised first aid at work course, which is usually now three days in total and is rewarded with a certificate upon completion. The certificate does run out and therefore it needs updating after 3 years (and every three years after that) to enable the person to continue as a first aider. These updates usually consist of refresher training which is often a couple of days in total. Qualified first aiders should be the first point of contact as opposed to an appointed person when an accident has occurred to ensure that help can be given quickly. First aiders are able to deal with basic first aid and are able to give cardio-pulmonary resuscitation and treatment for wounds, shock and other problems. Issues to consider when determining how many first aiders/appointed persons you may need:

- How many staff are on site at any one time?
- What are the types of accidents and injuries that could occur? Look at those that have occurred within the company or in industry in the past for some ideas.
- How many people are likely to be affected in an incident?
- Are there any specialised or high risk activities which might require an advanced level of first aid provision and possibly advanced training?
- How many buildings are there on site – would it take too long for someone to travel from one end of the site to another, for instance?
- What about non-regular activities?
- What about cover for those that are off sick, or on annual leave?
- Is shift-work a part of the business, or do people work out of hours?
- What about when first aid certificates run out –should there be an overlap so that people do not all have to go on refresher courses at the same time?
- What about people who are not office or site based, such as home workers, lone workers and those that do a lot of business travelling?

First Aid Kits

The contents of a first aid kit must be kept in a container as described in the Regulations, which means that they must be identifiable by a white cross on a green background. There is a standard list of items to put in it which is just for guidance and it really depends on the outcome of your risk assessment as to what you may need. Normally a variety of bandages, plasters, eye pads, dressings, disposable gloves (all individually wrapped to avoid spoilage and contamination) would be suitable as a minimum, for the average workplace. Often it is necessary to include an eye-wash, especially if there is no clean running water nearby to rinse the eyes.

No medication or opened items such as antiseptic creams should be kept in the first aid kit. Opened items could transmit infection, and no one is authorised to give out medication if they are not a medical practitioner. Remove any out of date items and keep kits well stocked. Regular checks should be made to ensure that it is well stocked, which could be done by a first aider or an appointed person.

Where will you keep the first aid kit(s)? They should be kept in a safe place away from contamination and damp, and as such they should have a designated place so that they can be accessed in a hurry. They should be kept shut.

How many will you need? Again this will depend on your risk assessment and the size of the building(s). They may well be kept near the first aiders, or they may be placed near the location of high-risk activities. You should have as many as you think you need, and you should review the provision regularly.

Depending on the risk assessment, smaller individual kits may need to be given to those who are not office-based or those who travel – they can then be kept in their vehicles or carried around with them. Remember to include these kits in the regular checks to see that they are fully stocked. Also, what about communication in the event of an incident?

How are these employees to contact you if the have been involved in an accident? Make sure that employees are told about the arrangements for first aid provision and consulted on the factors, such as the location of the first aid kits. Employees should be made aware of the staff who are first aiders and appointed persons and the differences between them, and how to contact them in an emergency. They should also be made aware of any procedures, such as whether all calls to the emergency services will go via reception, so that a consistent approach is taken and there is a reassurance that the correct directions will be given. This is just an idea and will not be appropriate in all circumstances. The main point is that *someone* calls for an ambulance in an emergency.

What about access for the emergency services when they do arrive – is it possible to get an ambulance in close to building? Will cars and business vehicles block their way, or is there a designated area for ambulances and fire engines?

Work at Height

Falls from height have consistently remained one of the biggest causes of workplace injuries and deaths over the years. Tighter regulation of this area which affects many employees at work has lead to the introduction in 2005 of The Work at Height Regulations, which apply to employers, employees, the self-employed, contractors and any person who has a degree of control over the work of others who may be working at height. The Work at Height Regulations 2005 give three different principles to follow, namely:

1) Undertake an assessment of the risks associated with all work at height and ensure that all work at height is planned, organised and carried out by competent persons

2) Follow the stated hierarchy for managing work at height activities, and

3) Do all that is necessary to prevent anyone from falling from height.

The Regulations have brought together all previous legislation regarding work at height and apply to all work at height whereby there is a risk of falling which is likely to cause injury, so in a sense that can also include standing on a step, or on the second rung of a ladder, and even includes places below ground level if the principle of being liable to personal injury applies. In fact, many injuries are likely to occur from what might be considered a 'low' height, such as standing on steps or on a stool, as perhaps people are less careful and work is less likely to be planned. Historically, construction workers (and others) were used to applying the 'two metre rule', which meant that any work involving being at heights of two metres or more would usually need a risk assessment of the work to be completed before work commenced, and extra measures put in place to prevent a fall.

Under the new Regulations, the two-metre rule has ceased to exist, and any work at height must follow the three principles above. It is now necessary, therefore, to put the same controls in place for work under two metres, as well as at two metres and above.

The hierarchy for correctly managing work at height is as follows:
1) In an ideal setting, to avoid work at height completely. An example would be for a window cleaner who changes to working from ground level with an extendable pole, rather than standing on a ladder to clean windows.
2) If work at height must be undertaken, use work equipment which prevents a fall, such as a mobile elevating work platform (MEWP). Be aware that MEWPs such as cherry pickers and vehicle-mounted booms need trained and competent persons to work on them, and they need to be appropriate for the job they are assigned to do. MEWPS can overturn if used on the wrong surface, or in high winds, and you need to be aware of any overhead hazards such as power lines that the extendable arm could come into contact with.
3) When it is not possible to eliminate the risk of a fall, use work equipment to minimise the distance and consequences should a fall occur, like using fall arrest equipment such as a harness.

As stated before, **all** work at height must be risk assessed and steps taken to reduce the risk of a fall **before** work commences. Measures which protect everyone, without the need for personal involvement or thought, should be put in place ahead of personal protective measures, which often rely on the individual remembering to use it, and use it correctly, and these methods may not be as reliable. So installing fixed guard-rails to prevent people falling from scaffolding, for example, would be preferable in the first instance to just giving everyone a safety harness, although the necessary control measures for each individual job will be determined by the risk assessment.

If you are an employee, or are working for someone else, you need to ensure that a relevant risk assessment has been undertaken to identify any hazards before you start work, and you must report any additional hazards that you come across. You must also follow the control measures stated, and use any equipment provided to you properly and in the correct manner. You should also have had the appropriate training and competency to ensure that you can carry out the job safely, and have been trained in how to avoid falling from height.

If you do any work at height, and indeed any construction work involving working platforms and means of access, I strongly suggest that you get yourself a copy of the Regulations and pay particular attention to the schedules which tell you what to do about inspecting equipment and the necessary reports and records required.

Sometimes, it may not be practical to do any work at height, and work may need to be postponed. For example, if the weather conditions could affect the safety of the operation, or the surrounding area where work is due to take place is unsafe (for example near open water, and where extra precautions may need to be included). Pay particular attention to any fragile surfaces, for example to materials that may be supporting a ladder, or to fragile roofing which workers may be working on or near, and ensure that workers and any others in the vicinity have been made aware of the dangers and put extra precautions in place such as using fall arrest equipment. In certain workplaces such as building sites, falling objects may be a problem and you should take steps to ensure that items are not thrown from height - in this instance a plastic chute could be used which safely takes items straight into a skip without workers trying to aim where they are trying to throw it. Do not store items near the edges of work platforms or roofing etc where it is likely that they could be knocked off the edge onto people below. Hard hats should be worn if anything is likely to fall onto people from above.

You will need to have plans in place to account for emergencies and the rescuing of persons who may have fallen from height. Employees should not generally be working alone if they are working at height.

Make sure that you inspect all fall arrest equipment and that there are regular documented checks. There should also be pre-user checks each time they are to be used. Employees using this type of equipment should be given training on how to inspect their kit for damages, and what to look for. Tiny cuts in lanyards can cause major deficiencies in strength, and mistreating equipment by leaving it out in adverse weather, or dragging it across the ground, or using it for another purpose such as a sling can cause it to become damaged. You must find out the type and frequency of the inspections to be carried out for different pieces of equipment, perhaps from the manufacturer, the Regulations and/or other health and safety guidance.

Ladders

Many of the accidents each year resulting from work at height are due to the use, or *mis*use of ladders. Ladders are often used for work activities that would be made safer by using an alternative bit of equipment. Ladders have come under a lot of scrutiny in recent years with much confusing information given by the media, which has often left employers and employees thinking that ladders are 'banned' and should never be used. It is correct to say that ladders should not automatically be used, easy option that they may seem, as under the new Regulations there is the requirement

91

always to assess the risks of any given situation requiring work at height and to use the most appropriate piece of equipment to keep the risk as low as practicable. If the work is deemed to be low-risk, and of short duration, then often a ladder may be used if the use of other work equipment cannot be justified.

Short duration is currently taken to mean work that will not exceed 15 to 30 minutes at a time. (That doesn't mean you can come down for a minute after the 30 minutes is up, and then do another 30 minutes and so on!) Think of it as a maximum of 30 minutes *in total*.

When using ladders remember the following:

- **Do not over-reach**.
  Always keep three points of contact with the ladder, i.e. two feet and one hand, which should stop any over-reaching and loss of balance. Always keep some rungs free above you, so that there is always something to hold on to.

- **Position the ladder correctly**
  It should measure one unit out for every four units up (a 75% angle from the vertical). Make sure that there is a good, even ground surface to avoid the ladder slipping at bottom, and a good level of surface that you are leaning it against to stop it slipping at the top. The ladder should be footed if possible which means having someone holding it steady at ground level. Stepladders need to be fully locked into place with a secure locking device.

- **Make sure that the ladder is suitable for use**
  There should be no missing rungs, the rungs should be clean and free from dirt and grease, and there should be rubber feet in place to give it some grip. Check that there is no visible damage to the ladder in any way. Wooden ladders need checking very regularly, especially if they are old as they can tend to disintegrate or be eaten away in places especially if they are kept outside. Don't cut ladders

if they are too long as the safety of the ladder could be compromised.

- **Inspect ladders regularly**
  Ensure each ladder can be easily identified from any others that you may have, perhaps by a number or a code - but don't use anything to mark it which could disrupt the normal workings of the ladder or enable it to become unsafe, as some paints can corrupt the composition of wood

- **Don't carry heavy weights**
  Don't carry anything heavy up the ladder with you – it can affect your balance and you probably won't be able to maintain three points of contact.

- **Dispose of ladders correctly**
  Have a system in place for disposing of your unsafe ladders safely, to ensure that they aren't left lying around, or that someone can't just pick them out of a skip and use them again.

Qu. Does that mean that I can't use a ladder to change a light bulb, once every couple of months?

You can, as long as the risk assessment identifies a ladder as the most suitable piece of work equipment to use, and the work will take less than 30 minutes to do. You also need to be trained in working at height. The ladder should also be positioned and used correctly to prevent you from falling.

Qu. How do I ensure that my employees (or anyone else working under my control) are working safely at height?

Make sure that you are a visual presence by supervising work at height – that way employees are likely to take control measures seriously, if they see that it is an important issue to you. Lead by example - if they see you using the correct equipment, or discussing the risk assessment with them

93

beforehand, they are more likely to see the management commitment to working safely at height, and will then understand its importance.

Make sure that you provide the correct equipment, training and work procedures to be followed and ensure that all employees working at height understand them and are able to use them correctly. Update any training regularly, perhaps with refresher training as well.

Keep coming back to the subject - toolbox talks are helpful as a regular reminder of the dangers and necessary control measures. If applicable, introduce permits to work for high risk activities. See the section on *Permits to Work* for more information on what this entails. Investigate all accidents or near misses regarding work at height to highlight any unforeseen hazards or behaviours which may need to be rectified to prevent a reoccurrence. Don't let any ladders belonging to other companies or (employees own ladders) be used as you cannot guarantee if they are safe for use. Make sure that the ladders you buy comply with the relevant British Standard for ladders.

Qu. How am I supposed to get on and off of a ladder correctly?

Ensure that the ladder is correctly positioned at a 75% angle, and then take steps to ensure it will not move — perhaps with someone footing it at the bottom for the duration of the work, or ideally having it securely tied it at the base and the top, making sure both stiles (the sides) are tied otherwise the ladder could swing and squash you against the vertical surface it is leaning against. Always leave at least 1 metre, or roughly three rungs clear at the top of where you want to get off, so that you have something to hold onto when you get off and on it.

Qu. What should a ladder inspection consist of?

As a guide, you should be looking for the following:
That there is no visible damage to the rungs or the stiles, with no broken or missing rungs, and that the ladder is not rotting or going rusty. See if

the feet are in a good condition, and that the ladder has not been tampered with or altered. See if it is wonky (as being left outside in the rain can warp some wood).

Are there any damaged tie rods, missing screws, items stuck to the feet, or any substances on the rungs? The ladder should be checked each time before use. Look at devising a written inspection sheet and then you can record these checks and prove that they have taken place.

The Disability Discrimination Act 1995 (DDA) (amended 2005)

It is illegal to discriminate against persons at work who have a disability. Under the Act, a person with a disability is described as someone who has a 'physical or mental impairment which has a substantial and long-term adverse effect on his [or her] ability to carry out normal day-to-day activities.' You will be unfairly discriminating against someone with a disability if you treat that person less favourably than you do others, in terms of their employment and also the activities and opportunities contained within that employment. Also, as a dutyholder you have a duty to make reasonable adjustments to the workplace or, in some cases, to how a work activity takes place. It is also, in many cases, against provisions of the Act to refuse to offer, or to deliberately not offer employment on the grounds of a disability. Seek legal advice if you think that you need clarification in this area. To make adjustments for an individual, you obviously need to know the nature of the disability and you will need input from the individual as to their needs. Company policies and procedures should reflect the notion of equality of opportunity.

Smoking

As of July 2007, nearly all enclosed and partially-enclosed workplaces must now be smoke-free. The new laws apply to all types of cigarettes, pipes and cigars. Every premise must have a specified no-smoking sign in place at the entrance. The law applies to complete buildings, i.e. those with a roof, ceiling and walls, and 'substantially enclosed' structures i.e. those that have a ceiling or roof, but have an opening in the walls that is less than half the total area of the walls.

Vehicles must also be smoke-free if members of the public are transported on them, and during paid or un-paid employment if the vehicle is used by more than one person (not necessarily at the same time). No-smoking signs will again need to be displayed within the vehicle, in a prominent position. If it is your own vehicle, you can of course do what you like in it unless the company declares otherwise on company business.

Obviously, you will need to prevent any smoking in the workplace, and any current smoking rooms and areas that fit the above descriptions of enclosed and partially enclosed structures. This includes temporary structures as well. You don't have to provide a smoking shelter (one of those things that look like a bus-stop, normally at the back of premises with several smokers huddled together to avoid the rain coming in sideways) but if you do, you need to ensure it doesn't become a substantially enclosed area, as above, and therefore it will be disallowed. It is a good idea to have a 'legal' smoking area outside, in a safe location, if you want to keep all the smokers in one place and stop them filtering into the road or other hazardous areas.

You might want to include the rules on non-smoking in the company health and safety policy, or include them elsewhere in your health and safety documentation.

'No Smoking' signs can be downloaded from www.smokefreeengland.co.uk, and further information can be obtained from the Department of Health, especially information on helping employees to give up smoking.

Construction

Key Legislation: The Construction (Design and Management) Regulations 2015. Commonly known as 'CDM.'

These Regulations came into effect on the 6th April 2015 and emphasise that safety and health is not an afterthought but instead something that should be thought about from the very beginning of a project. One of the main changes introduced under these regulations is that the client cannot now discharge its liability to a third person. The effect on business, especially SMEs, is that they will have to properly understand their responsibilities and the law relating to construction projects carried out for them or on their behalf as there will be no-one else to blame, or the ability to feign ignorance. The CDM Regulations should help to reduce the paperwork involved in a construction project and ensure that properly competent people are involved. Projects need to be notified to the HSE on Form F10 (available for download on the HSE website), if they last more than 30 days or 500 person days, so it depends on the length of time that the work will take, and the amount of people involved. See the Regulations for the full requirements that need to be met when undertaking construction projects. Please note that the CDM Co-ornidator (CDMC) was replaced by a 'Principal Designer' as an existing member of the design team. The principal designer has responsibility for coordinating health and safety during the pre-construction phase, which gives them the ability to influence the design from the very beginning.

Legionnaire's Disease

Legionnaire's Disease is a form of pneumonia which can affect anybody, but which usually affects the elderly and those that may not   easily be able to fight off infection. It is caused by the Legionella bacteria (hence the name) which is found naturally in areas of water such as lakes and rivers. It can also be found in man-made water systems such as air conditioning units and water-cooling towers and therefore could be found in the workplace. If you have a water-cooling tower you need to notify the local authority to tell them where it is. The Legionella bacteria can multiply in the right conditions, those being   where temperatures are warm (between

97

20C and 45C) and if there are other things present such as algae and rust which it thrives on. It likes stagnant water, so be aware of 'dead legs' in systems and the ends of pipes in a system where water can collect, and also showers and taps that are not frequently used.

People catch Legionnaire's Disease by breathing in water droplets containing the bacteria- it therefore cannot be seen or detected by the naked eye. It gives, amongst other things, flu-like symptoms and can lead to pneumonia. It is important that if an employee or a member of the public becomes ill, and you suspect that it may be legionnaire's disease contracted from the workplace, that the person sees their GP. If a diagnosis is made, it is important to inform the authorities as other people in the vicinity may be affected too, and also report it under RIDDOR as a work-related disease. It is relatively uncommon, and outbreaks are few and far between, but they do happen. It is not infectious and you therefore cannot catch it from someone else.

Measures that can be implemented to help prevent an outbreak of Legionnaire's Disease include the following:

- Ensure that water is kept at a high enough temperature to kill the bacteria
- Use a closed system for cooling water so that no droplets can get into the air, or use a dry air system air-conditioning unit.
- Treat the water with the correct treatments which help to kill the bacteria
- Don't allow stagnant water to accumulate
- Clean water towers effectively
- Appoint a responsible person with competent and sufficient knowledge of legionella bacteria and the necessary water treatment
- Have an action plan on how legionella bacteria and its risks will be controlled and keep records of any checks and tests that you make on the system
- Monitor the effectiveness of your control measures and your action plan.

It is important to:

- Do a risk assessment concerning legionella bacteria, highlighting the hazards and what could go wrong, and who could be affected (don't forget that members of the public may be at risk too if they are likely to be in the vicinity.) Look at whether water droplets are produced in a process that can subsequently spread or get out. Monitor your systems and processes, and review regularly especially with regard to control systems.

- How are you going to control the risk? For cooling towers and other similar objects, there should be a written scheme of maintenance in place with responsible people nominated and suitable precautions in place. The correct level of treatment should be used which you may need advice on. A consultancy or advisor who specialise in water treatment may be able to help you if you think you need advice, and can take samples of the water-although remember that the duty still rests with the dutyholder to ensure the health and safety of those who may be affected by the company's work activities. Although there is no specific legislation on Legionella there is an Approved Code of Practice which gives information on controlling the risks from Legionella bacteria. Duties apply however under the Health and Safety at Work Act and The Management of Health and Safety at Work Regulations.

Method statements

Method statements are documents that are produced to show how a job will be done, in what order the activities will take place, and the personnel and the equipment that will be used. They will often be used for one off jobs, or those which are a bit different to usual, perhaps because they are taking place in a different location or because they may involve hazardous work. They should use the information from a risk assessment for the work activity which has identified what might go wrong, and what precautions are needed to do the job safely. Method statements are often used if you have contractors working for you, or if you are working in conjunction with another company. Having a method statement helps to ensure that everyone has been involved in the hazard identification process, knows what is expected from them individually and is therefore singing from the same song sheet.

Bizarrely, I have seen many method statements produced before a risk assessment had been done - it should always be the other way around, as you can't write a working method procedure which shows how you are going to manage the hazards until you have identified the hazards, and that comes by doing a risk assessment! Method statements will mostly need to be done for each individual site or job that you undertake, as each site is different and the work will need to be done in a certain way for each job. In this case, a generic method statement may not be sufficient.

New and Expectant Mothers

A woman who is pregnant, or has recently been pregnant and/or is breast feeding (which can be for many months after the birth of her child) can be more susceptible to the dangers of the workplace. Some aspects of the workplace which might not normally cause her a problem could suddenly affect her health and that of her child.

As with everything, it is important to do a risk assessment which looks at any hazards and risks which may present themselves to this group of people. It is also important to do an individual assessment for each woman based on the requirements of her job and the hazardous situations which may present themselves. It is obviously necessary to consult with the individual as it is about her and she will know her job best (and also the aspects of her role with which she may have some discomfort in performing.) There is also a necessity to look at the risks associated with work activities for women who may be pregnant in the future, so it is a good idea to look at all work activities to see where there might be any potential hazards, for now and in the future.

Key areas to look for include:

- Work with or in the vicinity of chemicals and hazardous substances
- Work with or in the vicinity of excessive noise or vibration
- Manual handling
- Work at height
- Lone working
- Shift work
- Stress
- Cigarette smoke (although this shouldn't be so much of a problem with the smoking ban in enclosed places),
- The temperature (i.e. is it too hot or too cold)
- Sufficient lighting (to avoid eyestrain and prevent against slips, trips and falls)
- Standing for long periods (and also being in the same position for too long)
- The use of display screen equipment
- Exposure to radiation
- The possibility that violence and/or an aggressive situation may present itself

The findings of the risk assessment may mean that nothing needs to change, although in some circumstances it may mean that one of the following has to occur:

- That adjustments are made to their work, such as the hours/conditions/ nature of the work, or extra control measures are put in place to protect her
- Adjustments cannot be made, for whatever reason, and due to the risks involved you offer her alternative work whilst she is pregnant or breastfeeding

101

- In some rare cases she may, as a last resort, need to be put on paid suspension from work to protect her and her baby, although this is an area that you will need to seek legal advice on.

In looking to protect her from all the risks that present themselves at work, you should also include in your risk assessment any written health advice given by her doctor or midwife. Employees need to inform their employer, in writing, as soon as possible after they know they are pregnant so that procedures can be put in place to protect their health and safety and they should be encouraged to do so in good time. Women do not have to inform their employers, but then employers are not required to take action if they don't know about the pregnancy. You can ask to see the certificate from a doctor or midwife to prove that she is pregnant, and for confirmation of dates, etc.

Do ensure that frequent toilet breaks are allowed, and that there are facilities to rest/breastfeed/express milk. The toilets are not a suitable place for this! As a suggestion if you are short of space, you may find that the first aid room or other quiet rooms may be suitable.

There is no specific legislation relating to new and expectant mothers in the workplace, but the duties under the Management of Health and Safety at Work Regulations 1999 (as amended) and the Workplace (Health, Safety and Welfare) Regulations 1992 apply.

Violence

Violence is increasingly found to be a factor in some injuries received at work. It can incorporate abuse, threatening behaviour and/or genuine acts of physical violence.

It is necessary to report under RIDDOR any acts of physical violence which have happened at work, regardless of who perpetrated the violence. Violent behaviour can come from colleagues, other workers and non-employees such as members of the public (e.g. patients in a hospital). For the victim, being on the receiving end of violence can often result in injury, fear, stress and lost working days.

As an employer it is essential to identify where any possible displays of violent behaviour could come from and take action to prevent it as far as possible, and certainly manage any incidents to prevent a reoccurrence and deal with any ongoing problems. The reluctance historically to manage this issue in the workplace seems to have come from a fear of not knowing what to do about it.

Identifying violence in the workplace

As ever, it is important that you do a risk assessment of all work activities which could see employees be the victim of violence. This could include looking at:

- What are employees likely to come across in terms of violence? What are the probable situations where it could happen?
- Look at how employees could be harmed, and in what way. Without making assumptions, are they likely to come across angry people, or those with mental ill-health, or drunk people?
- Have there been any previous cases of violence? Ask employees, as you may not already know about them. Look in the accident book too, to see if any have been recorded in the past. You could think about devising anonymous questionnaires or surveys if necessary, if employees are worried about speaking up against colleagues. Get them to report even small cases or 'unimportant' events, as these events could be more serious if they were to happen next time. Have a clear format and process for employees to report acts of violence. Prove that it is an important area that you are taking measures to deal with, and that any acts of violence or verbal/physical abuse will not be tolerated.

- Identify *why* violence and violent behaviour might occur and any control measures you could put in place. Do your employees collect cash as payments, or do they work as debt recovery agents, in which case the motivation for violence could be theft? Don't make it easy for thieves to spot potential prey. Lock up premises and cars properly, and use signs to say that no cash is held on the premises or on people. Look at any industry information you can get about the potential for violent acts. As the dutyholder it is your duty to protect employees from events that might happen – so you must do all you can to prevent them happening. Are there certain times of the day that are more likely to attract violence, too?

  Have you thought about security issues? Could you make the workplace more secure? What about using cameras, not having people working on their own, especially at night, and making all areas well-lit with hiding places kept to a minimum. How can isolated employees have regular communication with others? If travelling and visiting employees are hard to keep a track off, what about introducing a buddy system whereby two or more employees make arrangements to call each other when leaving appointments or arriving home, to say that they are ok?

- Give training to employees to ensure that they feel competent to do their job and feel confident in handling difficult situations. They can be given information on how to not react to situations, how to spot aggressive behaviour and how to diffuse conflict. You should also give them information on the risks and hazards that might be found or that they might come across. You could produce a policy that gives guidance on what to do during an incident, and there are lots of tips on the internet on how to protect yourself at work which could be included, for example parking your car facing the exit of a location so that you can make a quick getaway.

- It is a good idea to have a policy on work-related violence which includes procedures for dealing with colleagues who commit forms of violence and the form of discipline that will take place.

In the event that an employee suffers from a form of violence, you should revisit your risk assessment and update it, with any new control measures that become apparent. The employee may benefit from talking to a counsellor after the event, and depending on the situation you should also possibly notify the police and get legal advice. It is important that all necessary steps are taken to prevent the employee being put in that situation again.

Young Persons (And the Working Time Regulations 1999 (as amended)

Key Legislation: The Working Time Regulations 1999 (as amended)
Children and young persons need to be protected from the hazards which are present at work.

A young person, in terms of health and safety law, is someone under the age of eighteen but above the minimum age at which they can leave school. A person will still be classed as a child if they are not over the minimum age at which they can leave school. A young worker cannot work more than eight hours a day, or forty hours a week, and must have two days per week off to rest, unless there are specific circumstances that require their attendance but you would be wise to obtain advice on this before you ask them to do more than their allotted hours. The law allows children of certain school ages to be able to temporarily, and for the purposes of work experience, take on the role of a 'young person' for that period of the work experience. Work experience is an important part of learning about what life will be like in the real 'working' world, and therefore if companies have put all the necessary precautions in place to protect the health and safety of children on work experience then they

105

should not fear taking on the responsibility. It is important to show children what work is like, and to give them an idea of the career path that they would like to follow. Schools will be very involved in the process and extra risk assessments will be required. Parents/guardians need to be made aware of the findings of any risks assessments before the work experience starts. When employing young persons, or taking on work experience children, you will need to think about, and protect against, the risks which may come about from their lack of experience, age, awareness and maturity. Young persons shouldn't be doing anything beyond their capability (physically or mentally) and neither should they be subjected to anything that could harm them, or any work that is deemed to be high risk.

Look at work activities which include hazardous substances, the likelihood of accidents occurring, noise, vibration, extreme weather conditions, work at height and physically demanding work, as these areas may not be appropriate, although they shouldn't necessarily prevent employment especially when young persons are in training. Ensure that young persons and those on work experience are always supervised, and are told the extent of the work that they can become involved with (and importantly that which they can't). There should always be a competent person in control and the risks must be reduced to as low a level as possible.

Ensure that the risk assessments are current and cover all work activities that the person may become involved in, and do use common sense as to what you believe the person may be capable of performing.

Remember that reaching a certain age does not always mean that a person has the maturity of thought and awareness that may be expected.

Slips and Trips

Slips, trips and falls account for a large proportion of all reportable injuries, perhaps because they can happen in any setting. Offices tend to have cables running along the ground, warehouses have boxes lying around on the floor, and motor vehicle repair workshops can have puddles of oil in the area, for example. Whilst slips and trips can cause injuries in themselves, such as twisted ankles and back problems, they can also lead to other, possibly more serious, accidents such as falls into dangerous machinery or down flights of stairs. It is important that a risk assessment is undertaken for slips, trips and falls in the workplace. It is possibly easiest to complete this by work activity, or by location. You could also have a policy on slip, trip and fall hazards which stipulates how they should be managed.

Firstly, try to identify the activities that could give rise to slips and trips. Trace the activity from the start to finish, encompassing all areas such as flooring, stairs, walkways, outside locations and carpeted areas and rugs. Look at the condition of the floor, and the objects that may be on the floor that could trip someone up. See if you can remove the hazard, such as the obstacle or trip hazard that is on the floor, or if it cannot be removed, control it to lower the risk to an acceptable level. As with the hierarchy of control in all risk assessments, PPE (in this case perhaps footwear) should be the last resort. Also take time to look at the surrounding areas, including the ceilings (which could be leaking water onto the floor, for example), and machinery in case of leaks and problems. Look at how people do the job – are they following working procedures correctly? Are they carrying items which obstruct their view? Are they wearing unsuitable shoes? Look at the environment – is the lighting sufficient so that people can view where they are going and see possible hazards in front of them? Is it an outdoor activity and therefore the weather may have an affect, such as making the floor wet or icy? What to look at, as a minimum, when doing a slips, trips and falls risk assessment:

- **The work activity**

  Has the work activity been organised properly so that it is performed in a safe way? How could it be improved? Look at the equipment used, to see if it has any trip hazards such as trailing

107

cables or parts that are sticking out. Are employees required to carry items that could obstruct their view, or unbalance them? Could the work activity be moved to a safer location, for example if it is outside, or if it requires walking up and down stairs whilst carrying objects.

- **Floor surfaces**

Are floor surfaces slippery, or are spillages often on the floor? How are they cleaned up and how are employees warned of their presence? Are there any openings in the building to the outside such as doorways or windows which could let in the elements? Could puddles, mud or articles from the work activity such as dust end up on the floor, or be walked in on the soles of shoes? What about snow, ice and frost outside, and the entrances to buildings which may accumulate these items outside the building. Look at installing a type of covered entrance over doorways and entrances to buildings, and mats to soak up water and other substances.

Are there any trip hazards such as doorplates, loose tiles or holes in the ground? Often layers on the floor such as rugs and carpet can slip against each other or curl up at the edges. Is the floor surface wearing out in places? Is the floor surface suitable for its purpose and the traffic that uses it, and it is anti-slip if appropriate? Does it need to be made rougher by using a coating, or does it need to be replaced? Talk to the manufacturer/supplier about your needs. Keep floors dry and free of debris and superfluous objects. Ensure that there is an effective cleaning regime in place

- **Footwear**

Are employees wearing the correct footwear for the work activity that they are performing? Open-toed shoes may not be suitable for jobs that involve carrying heavy objects. High heels may get stuck in grates, or cause twisted ankles or contribute to falls down stairs. The type of shoe worn at work may not be as much a problem in some industries and locations as others, but it may be an important factor in some. If you require employees to wear a certain type of shoe, or if you implement a non-slip shoe policy you may have to provide them for employees under the PPE Regulations.

- **Surroundings**

Are the surroundings conducive to preventing slips, trips and falls? For example, are there any dark areas which make it hard to see where you are going, or what is in front of you? Are there any obstacles in the area, or steep stairs? Look at installing good lighting, with ample switches and controls that can be found near the entrances to locations. Make sure that outside areas are adequately lit, perhaps with automatic lighting that comes on when movement is detected. Ensure that there is a good system in place for housekeeping, with designated areas for rubbish storage. Make sure you put wires away and stick cables down to prevent trips. Clean up spillages promptly and stick signs up when doing so, so that the cleaning does not present problems such as a wet floor. Don't leave 'warning' or 'wet floor' signs out all the time for no reason as people become complacent and start to ignore them. Try to perform general cleaning in quieter periods when there is less 'people' traffic and try to ensure that cleaners always leave a dry path through in case it is needed.

- **Other factors**

Could other factors contribute to the likelihood of slips, trips and falls? Look at work at height and manual handling, for example.

Could these types of activities induce these kinds of accidents? Are there any employees with special needs who may need consideration? Stairs cannot be used by everyone, for example those with balance problems. Is any PPE needed for work activities? Check that the footwear worn complies with any relevant safety standards and is suitable for the work activity and the conditions. There may be different condition to be met for shoes worn outside than those used inside on a dry floor. Do a bit of research into the types of footwear available.

It is important to give employees job-specific and general training on slips, trips and falls, and the hazards that they may come across and the control measures that have been put in place to lower the risk of someone tripping, slipping or falling. Get the point across that these types of accidents can be serious, and that they are relatively common. Give them the information that they need to identify any hazards, and what they could do to prevent them. Employees should know who they should report potential hazards to, and how they should do so. Decide who is responsible for each area in the workplace for cleaning and general housekeeping, and what is expected of them for ensuring that slips, trips and fall hazards are removed or controlled.

Make sure that visitors and contractors are made aware of your policy on slips and trips, and that any work undertaken by them on the premises must be in line with the policy. Think about how to incorporate checking for slips, trips and falls hazards as part of an inspection or regular walk around the workplace. It often helps to have a fresh pair of eyes to look for these (and other) hazards, as people can tend to miss things sometimes if they see them everyday. Keep records of any inspections and remedial action needed, and incorporate any findings into the risk assessment. Make sure that you record the assessment, if you have five or more employees, and revisit it regularly to see if it is still current, especially if there have been any significant changes or incidents.

Lone Working

People who work on their own, in areas where they are separated from others, or where there is no direct content with others can often be at risk because of the fact that there may be no-one in the immediate vicinity to help them should there be an accident or a problem. This also applies to employees who work different hours to the rest of the workforce or who work out-of-hours shifts, whereby there may be no access to the experience and skills of those that could help in an emergency. Examples of these types of workers include a person who works in a singled-manned refreshment van parked by the side of road, a shift-work technician, and a mobile worker such as a plumber.

The procedures that need to be in place will depend on the risk assessment – there is nothing to say that people can't work alone, as long as their personal safety is not in danger, and the risk assessment shows that there is nothing to prevent them from working alone, and the control measures in place are sufficient. It may be a different case if it is not deemed safe for the worker, and little or nothing can be put in place to reduce the risk. Pay attention to any specific legislation for certain industries which may impact upon lone working, especially high-risk activities.

Ask for information from other sites that your mobile workers work at, to ensure that their safety has been thought about in that site's risk assessments. Don't forget that you have a duty to protect the health, safety and welfare of your employees, even if they are working for someone else, or on another person's site.

There should still be some type of supervision and contact with lone workers, even if it is not direct. Employees should always be contactable and perhaps traceable by some means, be it by mobile phone, computer

111

device, satellite, or other means. A good way of keeping in contact that I have seen is whereby an employee logs in electronically to a central point, and has to make contact within a certain amount of time otherwise an alarm is triggered at head office and someone has to then call that person to check that they are ok. You can also get a similar procedure whereby the employee wears a device on their person which triggers an alarm at head office if no movement from the employee occurs for a specified amount of time. These types of lone-worker devices can give a fair share of false alarms, as you are often relying on an employee to remember to log in and out and to keep moving, but it is better to have a false alarm than an emergency that goes undetected. If you do think that you need to have this type of checking mechanism in place, then think about what you will do in the event of an alarm sounding. What if there is no answer when you call the employee on the telephone, or there is no telephone signal at their end? Will the next step be to send someone out to look for them? How will you keep in contact with the second person that attends them?

All lone-working activities should be well planned and thought out. It might be an idea to produce a method statement on the basis of your risk assessment, for each job, which would tell all involved how the job should be completed and the control measures in place to protect the safety of all involved.

Especially in the case of higher-risk activities, think about how you will rescue a lone worker in an emergency, for example if they became trapped in the area they were working in, or if they were to become the victim of violence. Should the work activity in fact be a two-man job? There may even be simple factors that would make the job easier if there was more than one employee, such as reducing the amount of weight that one person has to carry if the job requires the use of heavy equipment. What information, instruction and training is needed for your lone workers, and what level of competencies, experience and knowledge would you expect them to have? Lone working may not be a suitable form of working for anyone who is not deemed to have adequate maturity and hazard awareness, or who is not physically capable of meeting the demands of the job. A lack of understanding of the job might also make the employee panic if they are on their own, and perhaps make mistakes.

Home Working

It is becoming more popular these days for staff to work from home, either full or part-time or on an infrequent basis. Although they may be at home, they are still 'at work' whilst they are doing their job and it is therefore important that you take steps to ensure their safety whilst doing so. The HSE have the right to visit home-working employees at home, as they would do at their place of work. It can be difficult to work out how you will risk assess their home, but don't forget that you are only looking at how they could be affected whilst they are working – most people will have a study, home office or designated work area if they work at home frequently, and therefore it is predominantly this work area that you will be assessing. There is therefore no need for you to worry about the cat bed in the kitchen tripping someone up as that is none of your concern! Do a risk assessment of the employees' work activities, which may mean that you visit their home if necessary (and appropriate.) As with all employees who are at work, regardless of where that work is, you must undertake any additional risk assessments for young persons and new and expectant mothers.

When doing the risk assessment of the work activities, think about who else in the home could be harmed by those activities and the hazards found within them – there may be children and pets around and it is important that you and the employee are aware of the risks to the other inhabitants of the house and take steps to prevent them from coming to harm. Areas of concern to be risk assessed might include:

DSE – look at the use of computers at home, even if the employee is using their own computer or laptop.

Manual Handling – assess the loads that might be carried and the tasks that might involve manual handling.

Slips and Trips – look at cables from electrical equipment, and the housekeeping around the workstation or work area.

Electrical items – Ensure that the equipment that you provide is safe for use. Give the employee information on how to ensure that their own equipment is safe for use and maintained and inspected regularly. You are only required to provide PAT testing for the equipment that you as an employer supply, so you therefore do not have to check sockets and other fixed items.

Substances – risk assessments should be completed for all hazardous and dangerous substances used by the employee in their work.

Stress - Employees may miss the chatter and banter of the office and may find it hard to switch off and separate their work and home lives. Give them information on how to best deal with this, such as keeping a designated working area separate to living areas which is only used for work purposes. Ensure that they remember to take breaks, and that they keep in contact with work colleagues regularly.

It is probably best to record the individual risk assessment to prove that it has been done, and take the necessary control measures to reduce any risks. You could think about giving a checklist to the employee to regularly remind them to look for any hazards that may have cropped up. Make sure that you have given them the necessary information to enable them to know what the hazards might be, and what to look for. Most of the normal health and safety legislation applies to workers that may work for you at their home. Stress to home-working employees the importance of reporting problems and accidents which would otherwise go un-noted, and sort out a way of recording these in your accident book or other records.

Confined Spaces

Key legislation: The Confined Spaces Regulations 1997

A confined space can generally be defined as any totally or partially-enclosed space which, because of the fact that it is enclosed, creates a risk of serious injury or death to individuals or multiple persons. These serious consequences often occur because of a lack of oxygen or because of fumes from dangerous substances, along with other unsafe conditions. An example of a confined space might be a storage tank, some types of vessels, and an underground room with no ventilation. Often a confined space might not be dangerous in itself, but it is the work that is carried on inside it which creates the hazards, if gases are released, or fires happen. If the work activity includes the use of explosives, or hot work or the disturbing of residues which are hazardous to health within a confined space, then the risk may be very high indeed.

The Regulations dictate that the best option, and the option that should be the first choice, should be to avoid having to enter a confined space in the first place. If entry is unavoidable then you should devise *and follow* a safe system of work, whilst ensuring that emergency arrangements are in place to rescue those that are in the confined space should the need arise. Never undertake work in a confined space if it a risk assessment has not been made, and/or it is not deemed to be safe. As usual, determine and identify the hazards by doing a risk assessment of the activity ensuring that all aspects of the task are covered.

The Regulations also apply to self-employed for activities that you are doing yourself, and anyone else who is undertaking the activities on your behalf.

Key points to check when working in confined spaces, and doing a risk assessment:

- Make sure that you have appointed someone to check the safety of the work activity throughout its course, and only allow trained and competent people to do the work. Make sure that they have the correct equipment and PPE, such as breathing apparatus and harnesses if necessary.

- Devise and implement a safe system of work which is to be followed at all times. Do not attempt the work until rescue arrangements are in place for any emergency. Who will perform the rescue operation, and how will they do it? It may be different for every job. Do they have the necessary skills, competency, training and knowledge? What courses or training is it necessary for them to go on? Will you require them to perform practice rescues to keep the training fresh in their minds? What equipment will be needed to aid the rescue? How is this looked after and maintained?

- How will you enforce a 'no-entry policy' for confined spaces and raise awareness of what a confined space is, along with the procedure for who is allowed to enter one and in what circumstances? People tend to have pre-conceptions about what is and isn't a confined space, and you might find that you have employees doing jobs in locations that they are unaware of as being a confined space. For instance, many people think of a confined space as a small area or somewhere that they might feel cramped in, which is of course completely different to the definition in the Regulations. There are many ways to stop un-authorised access into confined spaces. First of all, you must work out where your confined spaces might be, that could give rise to serious injury or death because of what goes on inside it, and if appropriate mark them as such, so that people are made aware. Keep hatches, doors and openings locked with the key only available to trained staff who have a reason to have the key. Ensure that there is appropriate guarding and warning signs in place.

- Give training to employees on the hazards involved in working in confined spaces.

- Where possible, ensure that containers etc are emptied of any hazardous substances before entering, and use intrinsically safe equipment (non-sparking) where appropriate if there is a chance of igniting vapours etc. Make sure that there is a form of effective communication between those inside the confined space and those outside (i.e. the rescuers and helpers).

- It is important that, should something go wrong, no-one else should be put in danger, so ensure that there is the necessary equipment for rescuing and for resuscitation. Fatalities from work within confined spaces have included rescuers who have entered the confined space being unaware of the hazards and who subsequently are overcome by fumes as they haven't got the right equipment. Before you know it, you have more people to rescue and a huge problem on your hands.

- Make sure you think about ways of getting people out of confined spaces, especially if equipment has failed, or there is a casualty. If someone is unconscious, they become a dead weight which won't necessarily be able to fit through the small hole that they came in via, especially if the person is on a stretcher.

- Do you need to make confined spaces a Permit-to-work job? See the section on permits to work.

- In the event of an incident, will you need to shut down anything else (plant or processes) that may present further hazards at the same time?

- What specific information do you need to provide for the emergency services? Think about the information on the substance that the employee may have been subjected to, for instance.

- How will you raise the alarm in an emergency?

Permits-to-Work

A permits-to-work is a formal written procedure which ensures that all parts of a safe system of work are in place before work commences. It ensures that everyone knows their own responsibilities, and those of others. It also details who within the organisation has the authority to approve certain jobs and when, and then signs them off as completed at the end. Permits-to-work are normally used for hazardous work, such as jobs involving confined spaces, or for one-off jobs which, by the virtue of their length, numbers of people involved or level of complexity may potentially give rise to risks to health and/or safety. Many types of maintenance jobs follow a permit –to-work system. Permits-to-work are especially useful when new hazards have been found because a job or its location has changed, or the normal control measures need to be lessened to enable work to take place.

The permit to work should exist in conjunction with the risk assessment for the activity, and will determine how the job will be done, the hazards that have been identified and the procedures and control measures in place for safety.

If you find that it would be a good idea to use a permit-to-work system, start by detailing what jobs it needs to be used for and the limits to the authority for those with named responsibilities. They should not necessarily be used for everyday jobs, as the importance of them may start to wear off after a while.

Items to think about:

- What should the form look like? Preferably it should be in a logical format, with clear sections to show who has allowed the work to take place, and who has signed it off on completion.
- Will everyone understand the form and process? Make sure it is not too complicated or difficult to use.
- Who will be required to use it? Will this include contractors? What training will they need, in addition to your employees?

- What information needs to be given on emergencies, should one arise?
- For long jobs, what will you do about handing over responsibilities throughout on shifts changes or daily worker changes?
- What checks need to be made throughout to ensure that work is being done safely and as prescribed, and that control measures are working properly?

Electricity

Key legislation: The Electricity at Work Regulations 1989

Electricity can be deadly. Each year there are fatalities as a result of working with electricity whereby the necessary controls have not been in place.

Other outcomes of contact with electricity include shocks and burns, many of which can be debilitating and can dramatically affect the lives of those involved. Adverse events involving electricity often lead to other accidents occurring from the initial shock such as falls from height.

As always, a risk assessment should always be done before any work with electricity takes place, and all electrical equipment should be safe to use, regularly tested and in a good condition. The following points should be looked at as a minimum:

-Have the fixed installations of the business premises regularly checked, such as the electrical wiring in buildings. Have a Residual Current Device (RCD) integrated into the system, or put on individual sockets, to cut off the electrical supply in the event of a fault. RCDs need maintaining, too, as you need to regularly test them to ensure that they are working. Remember that RCDs don't protect against all types of electric shock, so they should not be the only control measure that you rely on to reduce the risk. Generally they should be rated at no more than 30mA otherwise they don't tend to do a great deal. Don't try to bypass RCDs as they are there for a reason. A competent electrician would be able to tell you how often a fixed installation check is required.

-Have portable appliances regularly tested (known as PAT testing) by a competent person (perhaps a contractor, such as a National Inspection Council for Electrical Installation Contracting (NICEIC) approved engineer, or someone from within your organisation with experience and who has perhaps trained on an approved training course). A portable appliance is generally something with a plug that can be moved around, and is not necessarily fixed in any one place. Examples of portable appliances would be kettles, laminators, fridges and heat guns. The HSE gives guidance on the frequency of testing that should be undertaken for a selection of items, although some companies choose to have most of their items tested annually. It is more important for those items which are regularly moved, that may wear out, or that may become damaged, to be tested than it is perhaps for a computer that is sitting permanently on a desk. PAT testing should occur along with frequent user checks, as anything can happen to an item within the year, especially wear and tear, and therefore it is best to check the item every time before it is used (i.e. you could look at the condition of its cables and plugs and check the unit for defects and wear and tear.) Training should be given to employees on what to look for, and it should be stressed that they should not be opening up the plugs or fiddling with the integral parts of the item as it is just a visual check to ensure that it looks in a decent condition. In addition, it may be necessary for regular inspections to be carried out on the portable appliance, which should be done by a competent trained person. Keep the results of and tests and inspections, and the frequency that they have occurred, to prove that it has been done. Work out how you will ensure that all relevant portable appliances will be included, and when the tests will take place.

-        Throughout the relevant tests and inspection completed on electrical items, check that condition of the insulation around conductors (e.g. wires) to ensure that the contents are properly protected and are not exposed.

- Ensure that the proper equipment is always used for the conditions present (for instance, waterproof equipment may be needed if it is to be used outside).

- Make sure that the people using the electrical equipment, or testing it, have the necessary competence, training and skills and that they have been told of the hazards and the findings of the risk assessment in relation to their jobs.

- Ensure that you know how to cut off the electrical supply in an emergency.

- Ensure that staff isolate electrical equipment before working on it.

- Look at the environment that the work is being done in – could the electrical equipment cause other hazards such as an explosion in confined spaces containing ignitable substances, or even damp places.

- Try to ensure that work is done at a lower voltage as is possible (regardless of electrical current type AC or DC)

- Give employees information on dealing with electric shock, such as putting up first aid posters which deal with this scenario.

- When working outside, watch out for any overhead and underground cables that you might come across – get information about their location and height above ground level before you start work, and measure the exact height of any equipment.

Pressure Systems

Key legislation: The Pressure Systems Safety Regulations 2000 and The Pressure Equipment Regulations 1999

A pressure system is a contained system that is fixed or transportable which holds liquid or gas under high pressure, like a boiler or a compressor. Pressure equipment is an item of equipment that does the same, such as an oxygen cylinder. Pressure systems and equipment needs to be looked after and maintained, as there can be serious consequences should something go wrong. Pressure systems can explode, releasing liquid and gas at high speeds, temperatures and pressures which in itself can cause injury (and damage to property) but also the impact of the explosion and being hit by other objects can cause major injuries. If flammables are contained within the pressure vessels then there is also the risk of fire.

There is a duty under the Regulations to take steps to prevent the failure of any pressure systems or equipment, and to ensure that it is safe to use.

Pressure systems should have, as a minimum:

- A safe system of work in place for using the equipment
- Trained operatives using it, who should be aware of the hazards and who should be able to recognise faults and possible problems. There should be warning signals in the event of a problem, and a list of the personnel who are allowed into the area to rectify the problem. Training should be updated and refreshed as appropriate.
- Make sure that employees know where the operating instructions are kept. What should take place in an emergency?
- A written scheme of examination and testing, at correct intervals as determined by a competent person. This should include any pipes, valves and hoses which make up the system. Look at the Regulations if you have any pressure systems on the premises (most premises will have a boiler or a compressor), as you will need to ensure that you have covered all of the requirements.
- Trained and competent person(s) to do any necessary repairs, maintenance, testing, and examinations.

Make sure that all pressure equipment is stored properly, or installed properly as appropriate. Check that it is suitable for the purpose with which you want to use it, and that the vessel or container is of the right material for the substance contained within it and it is labelled correctly for its contents. Check that safety valves are in place, which prevent the temperature and/or pressure getting too high by shutting the system down – and see if these can be easily defeated.

Asbestos

Key legislation: The Control of Asbestos Regulations 2012

These Regulations brought together previous legislation regarding asbestos management. Asbestos is a natural substance which over the years has been used in the manufacture of many items, such as cement, lagging for pipes, and insulation board to name but a few, due to it being a good insulator. There was widespread use of asbestos between the 1950s and 1980s, particularly in new builds and refurbishments. There are three different types, commonly known as blue, brown and white asbestos (crocidolite, amosite and chrysotile respectively.) White asbestos is generally thought to be the least dangerous, but in reality all three types can be extremely hazardous to health. When asbestos is disturbed it releases tiny fibres into the air that the naked eye cannot see. These fibres are long and hook-shaped and lodge themselves in the lungs and chest of the person who has breathed them in.

The main diseases associated with asbestos are mesothelioma (a cancer of the lungs and/or stomach), lung cancer and asbestosis (a scarring of the lungs). All of these conditions are made worse if you smoke. These conditions can take a while to show up – many years or decades, even - and there is no cure. The use of any type of asbestos including the use of second-hand asbestos-containing products is now prohibited. This applies to new use of the substance only - if asbestos is already in place and it is in a good condition then it just needs to be managed. The Regulations require that the dutyholder identifies where it is, maintains it if it is in a good condition and ensures it is not disturbed, and reviews it periodically to see if it is still in a good condition. This is called the 'duty to manage.'

123

You must also control any damaged materials to prevent release, and inform anyone working near it (such as plumbers and cleaners, for example) of its whereabouts, and the hazards it presents.

You don't *have* to do an asbestos survey but you do need to know where it is in the premises - look at the original building plans and any renovation paperwork if you are unsure or do not know the history of the premises, and if you know what you are looking for then have a look to see if anything looks like it contains asbestos. If it looks like it is, treat it as so for now until you are sure. Do not attempt to disturb it! Get expert help if you need it. Undertake a risk assessment to identify the hazards and to see what control measures and priority planning are needed. It is the duty holder's responsibility to ensure that the risk assessment is carried out to assess the risks and consider the condition of any asbestos in situe, or that they think to be there. Record and review as necessary (especially if more information comes to light.) Include the nature of possible exposure, what asbestos it is, any plans of where it is, and what needs to be done to manage the risks.

You must ensure that there is training for anyone who may come into contact with asbestos at work – and that is not just for those who are doing asbestos removal. Exposure should be prevented, or if this is impossible, it must be kept to as low a level as possible. It must be below the exposure limit of 0.1 fibres per cm over a time period as currently determined in the Regulations. You must also prevent any release of asbestos, and stop any people un-knowingly disturbing it in their work by giving them the necessary information about its whereabouts. If the asbestos is damaged and/or it is not in a good condition, there are three options for action. You can either:

-        remove it,
-        repair it, or
-        isolate it.

Most work involving asbestos removal must be undertaken by licensed removers (licensed by the HSE), with a few exceptions such as where the risk assessment shows there is no risk of exceeding the control limit exposure. There is a duty to notify the enforcing authority if work is licensable. However, since the 2006 Regulations were replaced by the 2012 legislation, there are now additional requirements for some non-licensed work to be notified, and written notes about the work kept as a record. The details should include as a minimum information about the workers involved and, where possible, the likely exposure. By April 2015, employees and the self-employed who undertake notifiable non-licensed work with asbestos must also be under a health surveillance scheme with a suitable doctor. The Regulations impose duties on owners/occupiers of buildings too as many people may have a duty to manage asbestos if they have responsibilities for maintaining the workplace or have any control over the work premises to any degree. It may mean that there are various peoples with duties over a premise and they should share the responsibility between them as applicable.

Medical surveillance must be provided for those working with asbestos, and air monitoring to check that they are not exposed to more than the limit. Suitable protective gear such as respiratory equipment should be worn by anyone working with it and ideally it should be controlled at source.

Asbestos-containing zones or areas should be identified by warning notices. Bags of asbestos-containing material should be clearly marked (and double bagged) to show that it contains asbestos – asbestos must not be classed as normal waste as it must be disposed of correctly.

Do not let employees or others take their work clothing home that they have directly worn whilst working with asbestos. I once heard of a woman who had contracted an asbestos- related condition years after repeatedly washing her husbands work clothes that he had been wearing whilst working with asbestos. Depending on the activity, the clothes may be asbestos waste or need specialist cleaning. You can, of course, get disposable suits to wear for this kind of work. The main point to remember is that, unless you are a licensed remover, you shouldn't be going anywhere near (or doing anything to) any asbestos which is not in a good condition.

Vibration

Key Legislation: The Control of Vibration at Work Regulations 2005

Regular exposure to vibration can cause a range of conditions known as Hand Arm Vibration Syndrome (HAVS) which includes vibration white finger (where the fingers 'blanche' and turn white as the blood vessels and nerves in the hand are affected) and carpal tunnel syndrome. At its worst it can be painful and debilitating, with numbness and pins and needle-like symptoms, and in severe cases the ability to write, grip and hold objects such as a cup can be affected.

Vibration travels from the vibrating object up through the arms, and in the case of whole body vibration can be transmitted through the body, for example, via the seat of a vehicle or a standing platform to give back pain - this is called whole-body vibration. The types of vehicle that may produce vibration are those such as excavators and forklift trucks. Whole body vibration is made worse by bumps, jolts and having no suspension. Hand arm vibration can come from hand-held tools such as breakers and angle grinders.

Recovery times from the symptoms of hand arm vibration get longer as the condition develops and the condition may become permanent if early action is not taken to prevent exposure to vibration. It is generally irreversible, and if you or employees are diagnosed with it you must be careful to ensure that it does not get any worse. In severe cases it has been found that people are not able to do their job if it involves further exposure to vibration, and it can also affect life outside of work, for example playing golf and DIY. It is very important to diagnose problems early to prevent them becoming more severe, and of course to have systems in place to prevent them in the first place. Things to look out for include tingling, numbness and pain in the hands (normally worse in the mornings, and in cold weather), the tips of fingers going white and loss of strength in the hands and wrist.

There are two values to look at for hand arm vibration - the exposure action value (EAV) and exposure limit value (ELV). The first gives a limit for daily exposure to vibration of 2.5m/s $A(8)$, which if surpassed you must take action to control exposure. The second value is the maximum amount of vibration that a person can be exposed to on any given day, which is 5 m/s $A(8)$ which should not be exceeded. Ensure that employees stay as low below both of these levels as they possibly can. Vibration calculators are available on the HSE website to help you work out where you stand. You can also use a device that measures vibration called an accelerometer to measure vibration levels – you may need some training on how to use it as they can be a bit complicated unless you are used to using things like that and working with mathematical equations.

What to look at, as a minimum, in the risk assessment:
- Identify all equipment which may pass on vibration - identify the jobs and activities where vibration equipment is used and the duration of use. Get information from the manufacturers who may provide you with general vibration data, but check its source and make sure that it is for your piece of kit, in the same conditions and on the same surfaces with which you use it.
- Work out who is at risk - different people may be more so, depending on the job and equipment that they use and how long they actually use it for.
- Put controls in place to eliminate the risk or reduce it to as low as reasonably practicable if above the EAV and reduce it immediately if above the ELV. Look to see if you can use a different piece of equipment instead

of the vibration equipment, or use a machine that does the work for you, such as a mini digger when digging up the road.

- Revisit the risk assessment regularly, especially if you introduce new equipment, there are any significant changes to the activity or people involved, or if you have a case of hand arm vibration diagnosed.

• Feed back into the risk assessment any results of health surveillance to see if control measures need to be changed or if more needs to be done.

- If necessary, prepare an action plan to deal with the worse cases, and/or the highest risk activities, first.

Health surveillance

It is stated under COSHH that health surveillance will be necessary for those who are considered to be at risk from vibration in certain activities. Health surveillance will help to identify any vibration -related conditions and ill-health hopefully at an early stage so that you can put measures in place to control it and make sure it doesn't get worse. It also provides checks that your control measures are working. You can also do basic questionnaires yourself to help determine any symptoms that employees have, and then refer them to your occupational health service provider (you can find these in the yellow pages), or you can get them to do it all via a programme of regular health surveillance. They will tell you if an employee is fit to work or not.

You should audit and monitor the service providers though to ensure that their service is working for you and check on their qualifications and expertise – they do really need to have experience of HAVS. You must report under RIDDOR if you are notified in writing that an employee has HAVS or carpal tunnel syndrome.

Make sure you keep individual records for each person under health surveillance- but this should not contain personal medical information.

Include in the record their normal exposure to vibration, and the outcomes of any health surveillance.

Health surveillance is measured in stages called 'tiers' of which tier 1 and 2 are questionnaires, and 3 to 5 are health assessments and/or a diagnosis by a qualified person such as an occupational health nurses or a doctor. As stated before, you may feel that you can do tier 1 and 2 in house, and then obviously use a provider for the rest of the tiers. This can be cheaper way of doing things, but you need to weigh up what will work best for you and if you have the competencies within the organisation. Training will probably be needed for who ever does it in house. Health surveillance is not really needed for whole body vibration — although you should question employees though about any back pain that they have which may be vibration -related.

You will need to act on the information you are given — you may need to give different employees with health problems different work, alternate their activity, or make sure that they use the equipment for less time. The findings of health surveillance may be given as anonymous information for groups of people. Employees are asked for their consent to you being given detailed information on their personal health and may say no, in which case you will still be given 'fit' or 'not fit for work' information for that person which you can then act upon.

Measures to put in place-
- Give information to employees on symptoms, best practice and how to use equipment properly, and how to report any symptoms. Encourage the reporting of early symptoms and have a procedure for doing so that everyone knows about.
- Buy dampened down equipment - ensure purchasers within the company know about HAVS and that they get information on the most suitable piece of kit.

- Ensure that equipment is in good working order – items such as grinders and breakers must not be blunt which makes employees in turn have to apply more pressure when using them. Follow the manufacturer's maintenance specifications and ensure that an action plan for maintenance is kept to. Reduce the need to grip and apply pressure by making sure the equipment is in good condition.

- Use the most suitable equipment for each job.

- Try to ensure job rotation so that one person is not using the equipment more than others, and they therefore get a break.

- Where possible, mechanise the process - get machines to do the job unaided where possible such as use attachments on JCBs, in breaking up roads for example.

- Ask who else is using your equipment i.e. contractors - consult with their employer on how to tackle and monitor HAVS issues.

- Supply gloves and other PPE which may help to keep employees warm. Normal gloves won't provide protection against vibration.

Noise

Key legislation: The Control of Noise at Work Regulations 2005

Exposure to loud noises, especially on a frequent basis, can damage the hearing, cause permanent hearing loss and give rise to symptoms of tinnitus which is an annoying and often constant ringing sound in the ears that can cause stress and loss of sleep. Even one-off loud noises can cause damage within the ear. Yet even if symptoms disappear after the exposure has taken place, it doesn't mean that lasting damage has not occurred.

Examples of noise-producers in the workplace include power tools, machinery, and also work processes such as hammering, and impact sounds. The Regulations introduced two lower limits for noise levels than were previously in place, with the new limits being 80 and 85 decibels (dB). If noise levels reach 80 dB then a risk assessment must be done to assess the risks to workers, and information on the risks and control measures must be given to employees. If noise levels reach 85 dB you must provide hearing protection and identify *and* label 'zones' to show the noise levels within certain areas. As with vibration, there is also an exposure limit value (ELV) of 87 dB, which must not be exceeded for any worker. Zones should be marked as such and should indicate where the wearing of hearing protection is compulsory.

What you should look at, as a minimum, when doing a risk assessment:

- The Regulations require that you assess the risks to employees and others from noise. Look at who may be affected and how, the control measures in place, what more needs to be put in place, and as usual review the risk assessment as and when necessary. Before you know what exposure is likely to occur, you first need to know the noise levels found in the workplace. To do this you can use a combination of testing equipment (either a sound level meter or a dosimeter which is worn by an employee to measure a personal exposure to noise) and you can look at the manufacturer's data of the noise-producing item. Where appropriate, you should ensure that hearing protection is provided and worn. Give information on the hazards, symptoms and control measures and the reasons for wearing them to employees. Health surveillance may also be needed, depending on the activities undertaken and the noise levels that are believed to occur. As with other health surveillance records must be kept and the authorities notified under RIDDOR if appropriate.
- Take account of any variations in noise levels, as they may be different on separate days.
- A competent person should undertake the risk assessment who is aware of the requirements of the Regulations.

Control measures to think about:

- A good general test to see if you might have a noise problem is if one person cannot hear another talking at a normal level when about one metre away, then you may be close to the exposure levels.

- Think about the impact of a combination of noises - even the radio can exceed the levels! Items like the radio when combined with machinery noise could well be a problem.

- If you have decided that hearing protection must worn, then take steps to ensure that it is - it may well be a disciplinary action if you think it necessary. Think about the consequences - what if someone comes back to you in twenty years with a claim for hearing loss because you didn't make them wear hearing protection- it will be your word against theirs if you have no way to prove that you did. Therefore, put systems in place and stick to them. Show how important it is by wearing them yourself and ensure that visitors do too (including health and safety inspectors – what a good way to impress them and show them that you enforce your policies by giving them a pair of ear defenders when approaching a hearing protection zone!)

- Buy equipment that is 'low-noise' and fit silencers or noise barriers on certain equipment and enclose loud machines to prevent the noise escaping.

- If necessary, look at implementing job rotation for employees to reduce exposure times

- Remember that maintenance may reduce vibrating panels and parts of equipment and worn parts which might give rise to noise. In the same way, ensure that machine parts are sufficiently lubricated.

- Some people find certain types of PPE uncomfortable. Provide some different types of hearing protection so that you can give employees a choice if possible, and see if they need to wear them with other types of PPE such as a hard hat or eye protection. The main point is that people must be as comfortable as possible otherwise they won't wear it. You could put well-stocked boxes of disposable ear plugs by entrances, so that there is no excuse! Ideally employees would be given their own personal ones with lockers or alternative storage areas to put them in to keep them clean. Make sure that any PPE complies with the relevant British Standard and that they are suitable for the job and expected noise levels. Watch they don't cut out too much noise, though, as this can be a hazard in itself if you can't hear people talking, or alarms going off. Hearing protection need only be worn when it is needed, and these times and reasons for using it should be explained to those that will be wearing it.

- Close doors to prevent noise escaping – ensure that the layout of the workplace is designed to prevent noise travelling, and that the main route through the building is not through a noisy area
- What about other people that you wouldn't necessarily think might be affected by noise – for instance, does the receptionist's desk back onto the machine shop?

As of April 2008 the Regulations and the lower action levels now also apply to the entertainment and music industries, including those workplaces where recorded or live music is played, such as pubs and nightclubs.

Occupational Road Risk

Many road accidents happen when someone is driving for the purposes of work. Just because a person has a licence it doesn't mean they are necessarily driving safely. The pressures of work, or complacency due to driving every day, can lead to unsafe driving in some cases.

A risk assessment should be undertaken to identify the hazards that employees may come across when driving on company business. A person will not necessarily be on company business whilst commuting to work, but this will include travel to other places within company time, and to places that are not the usual workplace. Consider all the types of vehicles that could be used on the road, including company cars and vans, and personal cars. Points to think about:

- Are vehicles maintained and kept in a good condition with regular inspections, services, MOTs and repairs to prevent failures of the vehicles? Do restraint systems work?

- Do employees have the correct license for the vehicle that they are driving?

- Potential hazards could include violence, work pressures, stress, tiredness and night driving. Also whether or not employees are regular or non- regular drivers, the length of the journey, the length of time that they have had their licence, speed limits, experience and the competency of the driver. Look at the possible weather conditions, which may also affect the load and sheeting on vehicles.

- Control measures could include advanced safe driver training, regular breaks, and capping the hours that can be driven each week.

- Encourage the reporting of accidents and near misses.

- What distractions could there be? What about the negative and positive aspects of satellite navigation and mobile phones? Ensure there are no extra hazards as a result of these devices.

- Could there be an alternative means of transport, such as taking a train?

- Give employees the information on hazards that they need, and any other information, such as the height of any large vehicles which may need to travel under barriers or bridges, for instance.

Consulting with Employees

Key legislation: Health and Safety (Consultation with Employees) Regulations 1996, Safety Representatives and Safety Committees Regulations 1977

There are two different types of safety representatives - appointed, official Union Safety Representatives if the organisation recognises that

134

particular Union, and elected Employee Safety Representatives who are chosen by the workforce. Safety Representatives should be consulted about any risks to do with health and safety concerning the work activities and the workplace, the control measures needed or those already in place, and the introduction of any changes that may affect health and safety (to equipment, shift work etc). They can play a positive role, if employers look to see it as a help rather than a hindrance. Safety Representatives are entitled to paid time off to attend training and to fulfil their role and you should give them the information that they may need to perform their role. You must ensure that everyone is consulted, whether represented by a Safety Representative or not. Safety Representatives' duties involve carrying out inspections, investigations, discussions, and being a part of health and safety committees and representing the views of the workforce. A safety representative should be made available during an inspection by the enforcing authorities to represent the views of employees, and they may speak alone to an Inspector if they so require.

You are required to set up a health and safety committee if two or more Union Safety representatives ask for one.

Information and Instruction
Information about health and safety can be given in a variety of ways, such as during health and safety meetings, one-on-one, via notice boards, during toolbox talks, or by putting memos in pigeon holes, for example. Often the best way is to use a variety of the above, or to work out which method works best for you. Job-specific information will usually need to be given before a work activity starts, as appropriate. Any new information may come about as a result of an accident investigation or because of the results of monitoring or health surveillance, for example, and because you need to update employees on changes, new hazards or control measures. For important issues, you may want to get confirmation that employees have

received the information by asking them to sign to say that they have read it and understood it, for example.

Human Factors or, 'man is not a machine'

We often think that accidents happen because of failures in systems, machines and other 'man-made' factors, but often it is the fallible, human part in a process that causes an omission or a problem. How a person reacts to a situation or performs an activity depends on their human characteristics, how they work, their individual behaviour, the environment and organisation that they work within and how the employee perceives themselves and others. Human factors, and ergonomics, suggest that by taking the person first, and then 'fitting' the job, workplace, setting, and environment around them, is the best way to go rather than doing it the other way round, as the human component is perhaps the hardest part to change. Looking at the human factors can reduce accidents happening.

- Consider the person – their ability to cope with pressure, their personality, their experience and knowledge, and their physical strength, for example.
- Consider the job – does it involve shift work (which can affect sleep, eating, home life, and social life), its instructions and physical requirements, and other factors such as are there inadequate rest periods or difficult processes, for example.
- Consider the equipment – look at its weight, location, and ease of use as a minimum.

For example, the emergency stop switch on a particular machine may be too high up for a short person to reach, which means that that person may stand on a stool and fall off it in the process of trying to reach the switch. They may also not be able to stop the machine in time in an emergency, or find increased stress at having to strain themselves to reach on a regular basis.

Consider human factors in your risk assessments and control measures, or when employees complain or incidents occur and you are looking for reasons why. The reason may be something really simple and easy to rectify such as adjusting the height of a chair, or using hands-free telephone if a person's neck hurts for example. Ask employees what works and what doesn't. Make sure that any changes are also assessed, to make sure that they don't introduce any new hazards. Also, having four different ways of doing something may confuse the person as they are not able to remember the instructions so keep things as simple as possible. There is no specific legislation for human factors, but it is contained within other Regulations such as the Manual Handling Regulations (i.e. looking at awkward shapes of loads and load weights) and The Display Screen Equipment Regulations (i.e. is the screen too close, is eye strain evident, is there limited room under the desk for legs to be comfortable etc.)

Work-related Stress

Stress is an unfavourable reaction to too much pressure. Work-related stress generally occurs when we feel that we are unable to cope and there is a loss of balance between what the job requires and our capability to do it. Its appearance can depend on how we perceive our jobs. Stress levels are different for everyone - one person may thrive on lots of pressure, whereas some people may feel that the same pressure is too much and they cannot cope with it. The job may need to be adapted to each person in some way to enable them to perform to their best.

We all need a little bit of stress, as small amounts are good for us as it helps to get the adrenalin going and aids motivation but too much can be bad for us. Too much stress can manifest itself in the body and lead to headaches, depression, and other more serious conditions. It can affect the ability to judge situations properly and therefore accidents can occur

137

because of this. It is necessary to manage the risks caused by stress by identifying them and controlling them. Undertake a risk assessment for work-related stress, with an action plan for taking things forward and a way of checking how it is going. It can be hard to know where to start, as stress can't necessarily been 'seen' and it can be hard to distinguish between other emotional issues, but perhaps start by looking at sickness rates and reasons for absence, to see if there are any patterns. You could give employees surveys or questionnaires (anonymous if necessary) to gauge their impressions and feelings about their work. You could even have a problem box for employees to raise anonymous issues.

Many factors can also bring on stress – changes (for example in work hours, job moves, travel time into work), the amount of control that someone has over their job, the demands put on them such as shift work, the culture of the workplace, and understanding where they sit in the organisation can all have an effect. Change can be a major factor – it depends on how it is managed and communicated, how secure employees feel, and the extent of the input they can have.

If you have a stress problem it can be handy to have support systems in place, such as a counsellor for employees to talk to, for work and other issues. There should also be procedures in place for reporting issues regarding stress.

It can be hard to distinguish where the stress originates from – it could come from home or work or elsewhere, or a combination. Positive features that you could introduce to help with preventing stress could be to help with work-life balance, maybe via providing flexible working hours.

The HSE have produced the Stress Management Standards which you might like to look at to help you identify and manage stress issues.

Vehicles

Vehicles used in the workplace can cause many nasty accidents, often because they are used in such close proximity to pedestrians or because they have not been driven or maintained properly. The commonest types of accidents are those involving vehicles that have turned over, those that have run people over, those that have pinned or crushed people against walls and other objects, and those that have dropped the load that they were carrying.

Forklift trucks

There are several different types of lift trucks, such as counterbalance, reach, seated, and standing.

Fork-lift trucks should be under a scheme of thorough examination and testing, which should also include the testing and inspection of its attachments such as chains and forks.

No-one should be driving a forklift truck unless they are competent, which usually means that they have passed a certified training course, which can be run at your premises which is often appropriate because employees will be training in the conditions in which they will be required to work in. They need to have basic competence training, along with job specific training to do with the tasks that they will be performing in the lift truck and then they should ideally have refresher training every few years (or less) to ensure their competency (and to get rid of any bad habits that have crept in). Any new lift truck drivers should be supervised on the job for as long as necessary. Ensure that employees have received training for the type of lift truck that they will be using, as the different types can be dissimilar to operate – most training courses and certificates can be converted to cover other types of lift trucks in short periods of time of further training. Keep records of any training given. If an employee comes from another employee and claims to have a forklift certificate make sure that you see evidence of this before you allow them to drive one.

If there is any chance that someone without the necessary competence and training will try to drive the lift truck, think about whether it is best to take the key out and only give it to nominated persons for access. Use the

139

restraining systems that are on forklift trucks — they are there to keep employees (and their arms and legs) in the vehicle should it overturn or be involved in an accident. Check that employees are using it by doing random checks. There are a few cases when it is not feasible to wear the seatbelt, for instance you might feel that it is not necessary and is an unnecessary hindrance as the forklift only goes at 2 miles an hour and travels only a few metres, and the driver gets off every other minute. Only your risk assessment can tell you if the risk is so low as to be negligible, but you should be prepared to have a good case to justify this decision if this is the stance that you want to take. If the lift truck is to be used outside, at speeds and/or the ground is not very flat, then I would suggest that you make sure that the seatbelt is always used.

Make sure that the correct attachment is always used for the task that is being undertaken — and only use something that is manufactured to be used for carrying persons if it is to carry people, such as a person cage. It goes without saying that standing on a pallet on the forks of a lift truck is not allowed...

Other vehicles

All employees driving any other type of vehicles must have the correct driving licence for that vehicle. For staff spending lots of time driving, you might want to think about providing some sort of advanced driving training to help them to drive safely. The section on Occupational Road Risk identifies some of the issues to think about for employees that spend time driving on the roads.

Vehicle Movements

Looking at how your vehicles move about in the workplace and restricting their movements to certain areas can be one way of ensuring that vehicles and people do not come into contact with each other. Try to limit the need for reversing, by putting in one- way systems, as this can help to prevent accidents involving the driver's blind spot. Look at the layout and try to make it as easy as possible for vehicles to manoeuvre within it, where they can be clearly seen by pedestrians. Segregate pedestrians from vehicles (and vehicles from plant) with clearly defined routes and areas for both. Avoid any blind corners. Introduce and enforce speed restrictions where necessary, and ensure that only trained and nominated drivers use vehicles. Check that floor surfaces are in a good condition, and that they are as level as possible and not slippery, especially when wet. Collaborate with any other companies who might share your site – you could perhaps do a joint risk assessment or at least share your findings with each other, and make sure that visitors and contractors obey the rules for vehicles on site. Make sure there is good lighting available at all times of the day that people may be working and that access and exits to roads are free from hazards. Put warning systems in place on vehicles (such as reversing alarms and horns). Only allow trained banksmen to direct vehicles but try not to use them unless absolutely necessary as they can get injured – they are, after all, pedestrians in a vehicle area.

Machinery

These days, all new machines should come with guards, but sometimes there can be a need to retro-fit guards to older machinery and plant. A risk assessment should be done on all machinery to identify the hazards, old or new, that may present themselves.

Common injuries include cuts, amputations, squashed or pinched fingers, broken limbs, and eye injuries which can occur from blades, rollers, heat, rotating parts and other dangerous parts that can pull you in, clamp down on you, or throw ejected parts at a person. It doesn't take a fool to realise that machinery can be very, very dangerous and therefore the correct controls need to be in place. The Provision and Use of Work Equipment Regulations (PUWER) go into much greater detail about the requirements to provide safe machinery and to protect against the risks from dangerous parts. Look at the hazards presented from all of the

141

machines that you use, including CNC (computer controlled) machines and manual machines. CNC machines are often thought to be safer, because they are mostly enclosed, but as we have discussed before, if the controls are defeated then they can be even more dangerous than a manually-operated machine. See what activities employees are doing in the vicinity of machines – be it setting them up, maintenance, cleaning, inspections or generally just working in the area. If it is a machine that could pull an object in, such as a rotating part, make sure that no loose-fitting clothing such as long sleeves and ties or long hair and jewellery is worn. There are many different types of guarding available, and the most suitable one should be used in each case to keep people away from dangerous bits of machinery.

Fixed guards – these are attached securely to the machine and need some sort of tool to remove them so that it makes it very hard for employees to take it off. Make sure that the tool is not left lying around so that it can be easily used, and ensure that it is put back after any maintenance.

Interlocking guards- they have a male and female counterpart which are placed together, and when they are forced apart the break cuts the power supply and the machine cuts out. Make sure that the cut-out time is not too long, as the machine may continue to perform its action for some time afterwards. As stated before, it is important that the interlock guards are not defeated.

Moveable guards- such as the 'fish tank' style of guard, which is often used if the machine parts are constantly needing to be moved, or are awkward to guard, and/or the worker needs to have some sort of access or a need to see what they are doing. Moveable guards should really only be used for lower risk machines. Most machines will have stop buttons and emergency stops, which should be tested regularly and kept clean. The 'hold-to-run' cycle should also not be defeated.

Training should be given to employees on the hazards associated with the machinery that they use – this should include, but not be limited to, the hazards, what guards to use and why they need to use them, what to do about defects and problems, safe usage of the machine, the PPE to be worn, other hazards such as contact with metal working fluids, how to isolate the machine, where the power supply is, the information contained within the written system of work, the findings of the risk assessment, who should perform inspections and operator checks, and the nominated users.

Metalworking fluids (coolant) provide lubrication for machine parts and help to keep machinery cool when it is in use. Metalworking fluids should not be breathed in and should not come into regular contact with the skin as, in certain situations, this can lead to the development of dermatitis and other conditions such as occupational asthma. It is often necessary for employees to be under a scheme of health surveillance if in regular contact with metalworking fluids. Metalworking fluids should be controlled and tested to check that they are at the correct concentration, consistency and pH level, and that excessive bacteria has not been able to accumulate. Those that do the checks should know when to change the metalworking fluid.

It is important to minimise the escape of any metalworking fluids via mists or vapours, for example in open systems or if machine doors are kept open. There should be no eating or drinking around metalworking fluids, or storing dirty rags with the substance on in pockets where it could come into contact with the skin.

Health and Safety Training

Health and safety training is given primarily to ensure the health, safety and welfare of all involved in the workplace. It can consist of on-the-job training or 'classroom' training on general aspects of health and safety.

143

You should provide induction training for all new employees, along with job-specific training for health and safety. Extra training should be given when there are updates in the law and its requirements, or changes in the company's health and safety policy and risk assessments, amongst other things. Think about the training you may need to give contractors and others that may work for you. Where will you keep the training records? Who will identify the training needs for employees?

For general health and safety training, you can get people to come to you in- house, or you can send employees on courses. A good start would be to look at the IOSH courses (Institution of Occupational Safety and Health) or those of the CIEH (Chartered Institute of Environmental Health) which is the awarding body for a variety of different health, safety and environmental courses.

Recruitment

Depending on the size of your business, you may be able to do everything to comply with health and safety requirements yourself, along with the help of one or more employees or others as competent persons. If you need more help, and are perhaps looking to fill a particular health and safety role, it is perhaps best if you recruit someone with specific health and safety knowledge and experience and more often than not, a qualification to prove it. A good start would be a NEBOSH certificate, which then leads onto a Diploma at the higher level. NEBOSH stands for:

The **National Examination Board in Occupational Safety and Health**

and is an independent, established body with courses held all over the country by various different course providers.

NVQ level 4 in Occupational Health and Safety is another good one, although there are also degrees and post graduate qualifications at the higher level. Look not just for awareness of the law or an ability to spout legislation at ease, but the ability to solve problems logically, to get on with the job, and a general positive attitude and willingness to do the right thing for other people, even if those other people don't see it that way! Health and Safety isn't for the faint hearted, that's for sure. Define the key responsibilities that the job entails and what you are looking for, to make sure you get the right person.

Emergency Procedures

Most businesses need to have emergency procedures in place to get people out in good time, especially in the event of a fire or an escape or release of a hazardous substance.

How will you get everyone out in an emergency? Has a specific risk assessment been done for fire? Test your planned routes of escape and feed back any problems into the risk assessment. Look at any persons who may need help in exiting the building.

Depending on the nature of your business and its location, you may need to think about other emergencies that could occur, such as an explosion, or a bomb threat. Try to cover as many different scenarios in your procedures, as far as possible.

Contractors

A contractor is basically anyone who is brought in to work at your premises that is not an employee. The 'client' is technically the employee who uses the contractor in their workplace. There is a great deal more information, especially for notifiable construction work, in the Construction Regulations (CDM) on the definitions of contractors and clients, and their specific responsibilities. In notifiable construction work the client and the contractor will have specific legal responsibilities that differ for those concerning normal work activities.

You don't want anyone on your premises if they are going to work unsafely, or in a way which could give rise to an incident. So how do you know if they are competent and work safely? Firstly, find out how they are going to do the work – look at their method statements, safe working procedures and risk assessments, and whether a permit to work may be required.

Consider also their qualifications, knowledge and experience, and the level of supervision that management will provide. You need to give them an induction for your site (including emergency procedures, fire exits, where to go and who to talk to if there is a problem etc). Health and safety requirements that they must fulfil should probably go into the contract that you have with them. Remember that you cannot transfer your duty or liability to anyone else, though. Work out how communication will happen between all parties, and if sub-contractors are involved, then this becomes even more important as no one must be left out of the loop. You can also ask about their health and safety performance before you work with a certain contractor- including their recent accidents, and any failures of equipment etc. You can check on the HSE website under Prosecutions and Notices too as any convictions and Notices are listed on there. Ask if they belong to a professional trade organisation or body, and see if they perhaps have their employees use a health and safety passport scheme (which includes basic health and safety training that applies to all sites, with regular refreshers that are often needed to ensure it is current.)

What information do they ask for from any sub-contractors that they use? Consider the risk assessments that need to be done with all parties involved. Have regular meetings that every contractor is present at- you can consider any changes that need to be made to policies and procedures, and give feedback on performance. Try to get several quotes before you use a contractor and ask for all the information that you want to see upfront. If people are less than forthcoming, be wary, as they should have nothing to hide if they are a reputable company.

Even if you regularly use the same approved contractors, always try to undergo the same procedure each time with a new job where possible.

Inspections

Regular inspections of the workplace can help to identify any hazards that have not been previously detected. It is a good idea to start by covering the topics included in this book, and anything other subjects and legislation that is relevant to your business. Your risk assessments and any other findings that you have will also help you to know the areas of your business that may need monitoring and regular inspections to check that your control measures are working. Industry guidance can also help with the specifics for your industry. Get a plan together of how you are going to do regular inspections, get into the habit of doing them, and ensure that directors and senior management are on board. Have a walk around and see what you can find. If it looks wrong, it probably is.

****

Chapter 5

Investigating and Reporting Accidents

Key Legislation: The Reporting of Injuries, Diseases and Dangerous Occurrences Regulations 2013 (RIDDOR)

Certain types of accidents and injuries that happen to employees (and sometimes others) need to be reported to the HSE, so that the authorities can compile data and statistics on accidents, so that they can help and advise in situations to prevent a re-occurrence, and so that, in many cases, they can investigate to establish what has happened.

What to notify and when:

- A **Death** (of an employee, the self-employed and any others on the premises)
- A **Specified Injury** – for the above people – including injuries such as amputations, loss of sight, unconsciousness, and stays in hospital for 24 hours or more.
- **Over Seven-Day Injuries** – that happen to employees and are not classed as a major injury but happen when an injury is sustained so that they can't do their normal job for seven days or more, not counting the day of the accident itself but including any weekends or annual leave. This may include injuries such as back injuries, violence etc. They should also be reported within fifteen days of the injury occurring. You no longer need to report over three day injuries but you should still keep a record of them

- **Some Dangerous occurrences** – This is where something happens which doesn't result in an injury or harm, but could have. It is also sometimes known as a 'near miss'.

- **Some diseases and work related ill-health**– certain work-related diseases need to be notified to the HSE but only if they have been identified by a doctor in writing, i.e. on a medical certificate, and if the work activity is one of those listed in the Regulations or it is a condition that is likely to occur from the work activity. This includes conditions such as occupational dermatitis, occupational asthma, and Hand Arm Vibration Syndrome. You should notify the HSE as soon as you know about the disease or condition.

Who you should notify:
All reporting can be completed via the HSE website (www.hse.gov.uk)

Self-employed
If you are self-employed you should also report under RIDDOR if there are any deaths or dangerous occurrences or you or someone else is injured on your premises, or if your doctor has diagnosed you with one of the listed work-related conditions. But, if you are working for someone else on their premises and you have a reportable injury, you need to tell *them* about it as it is *their* duty to report it. If in doubt, you can always report it yourself.

Question. Who within my company should make the reports under RIDDOR?

It doesn't really matter as long as it is reported –you may decide that HR or someone who deals with paperwork would be best, or the health and safety dept if you have one – it is however probably best to have a set procedure in place to keep it consistent so that no accidents or cases of ill-health are missed.

When Things Go Wrong

Unfortunately, sometimes things do go wrong and accidents do happen. That's not to say that that's ok- it's just a fact of life. The important thing is to find out what went wrong, and put it right so that it never happens again. It can be a very stressful and upsetting time for all involved – there is the family of the injured or deceased person who will want to know why their loved one has been hurt, and will naturally often want to know who is individually to blame. There may be witnesses who are traumatized or who feel guilty that they didn't do enough at the time or that it 'should have happened to them.' There may be guilt and worry at the management level of the company who wonder if their neck might be on the line. No matter who is involved, the aftermath of an accident, be it small or serious, is a difficult time for all, especially once the enforcing authorities get involved. It is important to take the time to reflect and review what happened, and then take steps to learn from the mistakes made and ensure they don't happen again.

Depending on the circumstance of the incident, it is possible that the person injured or harmed (or their representatives) may take a civil claim against the company, and the Enforcing Authorities may also prosecute if there has been a breach of law. If either of the above happen, then you will need to get professional legal advice. Health and Safety law and the legal system is an expansive subject requiring extensive detail beyond the scope of this book, but I have tried in the following sections to give a brief outline as to what might happen in an investigation by the HSE, and then also what happens during the process of being taken to court for a breach of health and safety law. This is intended solely to give an outline of what can happen – each and every case is unique and has its differences based on the circumstances of what is alleged, and therefore the information in these sections should be taken as a guideline as to what can happen. Your legal advisor will be able to give you any relevant information that is appropriate to your circumstances.

150

If you have an accident, it is important to ensure that no-one else in the area can be hurt, which may mean that you need to cordon the area off or somehow make the area safe. Follow your emergency procedures for the type of accident that it is, and ensure that a first aider (or another person) has notified the relevant emergency service. If it is safe to do so, you may need to leave the scene as it is for any ensuing investigation by the authorities.

If in doubt, call the Police or the HSE Infoline, informing them of what has happened, and follow guidance from them as to what actions to take next. Make sure that enough information has been given to those in the vicinity about what has happened, and any continuing hazards. If there is any evidence that may disappear naturally, for example substances and spillages that may drain away in time, it might help to take photographs to show a representation of the scene if deemed necessary.

Accident Investigation (including near misses/dangerous occurrences and ill health)

All accidents, major or minor, need to be investigated for several reasons. Firstly, you have a legal duty to find out what went wrong and to ensure it doesn't happen again by putting the necessary control measures and precautions in place. Secondly, undertaking an accident investigation is a useful aid to help an organization gain a greater understanding of the risks involved in its work activities, and to point out areas which may need review and improvement. Thirdly, the HSE will often ask for your findings of an investigation after an incident, and it looks very good to have prepared your own report, with an action plan and control measures identified for taking things forward.

Definitions of 'near misses' and 'dangerous occurrences' vary, but generally a dangerous occurrence is when something unplanned happens which *could* result in an accident or injury, but which doesn't happen this time. For example, a shelving unit collapses but nobody was standing near it and therefore no one got hurt. Even though you don't want them to

151

ever occur in the first place, a dangerous occurrence is a useful tool to learn from – it highlights what *might* go wrong at some point, and therefore you can prevent it from happening again in the future and actually causing the harm.

Dangerous occurrences and near misses are similar, although you could differentiate between them by saying that a near miss is something which didn't actually happen, but could have. For example, a broken rung in a ladder which hasn't been used yet might be identified as a near miss. The 'unplanned event' has not yet been realised, let alone the harm, but there is still the potential for harm in the future, should the ladder be used. In reality they are much the same thing and there is nothing wrong with putting them in the same bracket and indeed some people don't differentiate between the two. Obviously a dangerous occurrence has special meaning under the RIDDOR Regulations and an incident should be reported if it comes under the prescribed list. Cases of ill health will normally be flagged up by a medical practitioner during medical or health surveillance, or if a person has been to see their doctor and produces a doctor's note which states that they have been diagnosed with a work-related disease or condition. The reason for the ill health must be investigated to ensure that control measures are working, and to identify any failures, and to see what else needs to be done to protect that person and others in the future. When recording your investigation it doesn't really matter how you do it, as long as you get the required information out of it. The following is a suggested list of questions requiring answers that you may want to include in your investigation:-

- What was the date of the incident and the time that it occurred? (In cases of ill-health, there may be a range of dates between which the person was working with a substance, for example, or a range of dates for which the ill-health existed. Try to identify when the ill-health may have started.)

- Is the incident reportable under RIDDOR? (See the section entitled *Reporting of Accidents to the Authorities*)
- Who was injured, and how? Where were they injured, and was anybody else involved?
- What activity was taking place – was there anything unusual about the day or the behaviour of the staff involved in the incident? What was the weather like, and did it affect the activity taking place? What about the time of day? What were the working conditions like?
- Look at the personnel involved – was it their normal job role and activity, and were they trained, competent and suitable for the job?
- Are there any different opinions on how the incident happened – what are the reasons for these differing views? One person might have been in a better viewing position than another. Check that no assumptions have been made by asking people to go through their account of the incident in great detail, from start to finish.
- What caused the accident or ill health– was there an exact object or substance which directly or indirectly caused the incident to occur?
- Look at the relevant risk assessment – had this identified the hazard previous to the accident, if so, what went wrong? Were control measures not used, or were they defeated or ignored?
- What written or un-written procedures were in place to determine how the work activity should be done? If there weren't any, why not? Ask the relevant manager or supervisor how safe working procedures are communicated to staff.
- Did the workplace affect the situation – was it a mobile workplace, or were employees working at other persons workplace?
- Was there anything about the work activity that made it hard for the employee to do the job? For example, was it a difficult activity requiring excessive mental or physical effort or repetitive actions?

153

Were the materials involved very heavy? Was there a lot of noise on site and the employee couldn't hear his colleagues when they were shouting at him? Is English not the employee's first language, and she didn't understand what was expected of her?

- What about safety signs and notices? Have they generally been ignored? Have they fallen down, or are they not relevant anymore?
- Was the correct equipment used for job, including protective safety gear and PPE?
- Was health surveillance and/or monitoring in place? Did it highlight any similar issues previous to the incident?
- Has the incident happened before? Were any steps taken to ensure there would not be a repeated incident?
- What legislation is applicable? Where does the company stand with regards to compliance with this legislation?

There will be other investigative routes that you may follow, especially as some answers may lead to other questions that need to be answered. This list should hopefully help you to make a start in the right direction though. Do get any Safety Representatives involved as they can often help with the investigation (although they may choose to do their own). Employees generally have the knowledge and experience that can add to the depth and maturity of the investigation, and they are often the best people to consult when looking to introduce new control measures as they know what will work and what will not.

If possible, try to come to some sort of conclusion as to the underlying cause behind the incident – trace it back a step at a time. For example, let's say you are undertaking an investigation as to why an employee cut his hand when using a knife to remove wire ties from a package. We know that the blade of the knife cut his hand, and then on talking to the injured employee and his colleagues it becomes apparent that he did not know that it was company policy not to use knifes to cut wires on packaging as it is deemed dangerous. When questioning the employee's line manager, you realise that the employee hadn't received training on opening packages, as a training need had not identified by management. The underlying cause may therefore be attributed to managers not realising that there was a training requirement for unwrapping packages correctly and safely, and for ensuring employees are trained in the use of knives and their limited use in work activities. The cause of an accident is rarely the immediate action that happened, in this instance the blade cutting the employee's hand– it can often be traced back to a management failure in some respect.

Witness statements

The Authorities may take statements from those involved in, or witnesses to, an accident but it can also be a useful tool to help with your investigation if you wish to take your own. In some circumstances it may be necessary to take them to aid the investigation and work out why things happened, for example in lengthy or complex investigations where information may be lost over time, or where there are conflicting opinions as to how an accident occurred. Writing it down can also prevent important facts being lost. It may be necessary to wait to take the statement, in case of shock or if someone is taken to hospital. Ensure you take it in a quiet setting, and let someone be with the witness if necessary. The reason for taking a statement is not to apportion blame on anyone, it is just to find out why the accident happened and it is important to tell this to the person you want to take a statement from. Anyway, people tend not tell the truth if they think they will get into trouble and a statement full of untruths is useless to your investigation. It is often easiest for the person who asks the questions to also write the answers down (but importantly in the witness's own words) to get the facts into a logical order and to ensure that nothing is forgotten.

Having the questions ready before you start should help with getting the information that you need from the person, although if the person you are taking the statement from deviates from your line of questioning, do not worry as long as the information they are giving is relevant to the investigation.

The following is a list of example generic prompts that could be used when taking a statement. As with the questions used for the investigation, it will be necessary for you to add your own questions which are relevant to your individual incident:

1) Their name, age, job title and relationship to the injured person.

2) What they saw, heard, smelt or felt during or immediately after the incident.

3) A list of the chronological events of what happened leading up to and during the incident.

4) Their knowledge of the incident, and where they were placed in relation to the event and what they were doing at the time.

5) The environmental conditions at the time, including the weather - did this affect their sight or their ability to hear what was going on?

6) Their knowledge of the condition of any plant, machinery and/or controls involved in the incident.

7) Were there any relevant issues before the incident occurred that are related to the incident?

8) Their knowledge of the way things are 'normally done' – this may include knowledge of cutting corners, or doing things in a different way than management may expect they are done. Is the way things are normally done consistent with company policy?

9) Why and how they think the accident happened.

10) What health and safety procedures are in place that they know of? Are they aware of the findings of the relevant risk assessment?

11) Information on any injuries sustained by themselves or others that they know of.

The conditions after the incident and any action taken by them or others that they know of.

13) Any other information that they may think is relevant.

Ask the person to sign their statement, perhaps with a declaration to say they have freely given the information contained within it and that the information given is true to the best of their knowledge and belief.

Enforcement

During an inspection an Inspector may give one of the following where he or she sees fit and in the event of a breach of health and safety law. The type of enforcement will obviously depend on the breach and other factors such as whether previous advice has been given on the same matter, and the severity of the hazards and the resultant risk arising from the situation.

Verbal/written information

For less serious issues, the Inspector may tell you during the inspection the issues that need to be put right, or they may write a letter, often asking for you to write back within a certain period of time to say what action you have taken. Dependant on various factors, they may or may not pay a follow-up visit.

Improvement Notice

An Improvement Notice may be issued when certain aspects of health and safety within a company are not being controlled properly and health and safety legislation is being breached. A time period for rectifying the situation is given, ranging from a few weeks to a few months, depending on the situation in question and how long it may take to rectify. At least 21 days however will be given to enable time for you to appeal to an industrial tribunal, should you wish to. They are often accompanied by a Schedule which tells you what you need to do to put the matter right. The Improvement Notice may be given at the time of the visit, or it may be sent in the post, but you will be told that you will be receiving it before you get it. A notice that has not been complied with often results in a prosecution. It is possible to apply for an extension to the time limit given, if you have a valid reason, but do make sure that you apply in good time, and don't let your date of compliance come and go without doing anything about it. The Inspector will generally come back to check that the actions have been taken, and the situation has been rectified, or they may ask for proof of your actions taken such as a letter from you detailing what has been done.

Prohibition Notice

A Prohibition Notice may be issued when the Inspector considers that there is a risk of serious personal danger or imminent danger. It can be issued for work which has not yet started. There are two different types of Prohibition Notice: Immediate and Deferred. An Immediate Prohibition Notice means that work must cease immediately until matters have been rectified accordingly. A Deferred Prohibition Notice will state that the work activity must stop after a set time period -for example if it would be unsafe to stop work immediately, then the Inspector may allow some time to make the situation right before stopping work. In most cases Prohibition Notices are given on the spot. They may also exist in conjunction with an Improvement Notice. As you will be aware, it can be very costly to stop production- so don't let it happen!

If you want to appeal either an Improvement or Prohibition Notice, look at the information that is given to you with the service of the Notice.

All types of Notices are listed on the HSE website for all to see, along with prosecutions taken.

****

Chapter 6

Health and Safety Prosecutions

Civil claims can arise when a person injured at work claims compensation for their injury or ill health from their employer. Civil claims are generally based on a duty of care – i.e. everyone has a duty of care to everyone else and the employer may have broken their duty of care and that negligence therefore resulted in, or contributed to, the injury to the employee.

Personal injury compensation may be covered by Employers Liability Insurance. There is generally no involvement from the enforcing authorities in civil claims unless they decide to prosecute for their own reasons.

As soon as you become aware of a civil claim you should seek legal advice. The Law Society regulates Solicitors and may be able to put you in touch with a suitable legal advisor.

Health and Safety prosecutions

Prosecutions are normally undertaken by HSE inspectors/ EHOs in the first instance in the Magistrates Court, and by a solicitor or barrister in the Crown Court. There are various factors to satisfy in order for a prosecution to be taken, such as there has got to be a good chance of a conviction, and it must be in the public interest to take the prosecution. The company or individual prosecuted will be the defendant, and will be alleged to be guilty of an *offence*. The different types of offences and the places at where they are heard are as follows:

Summary offences

These are only heard in the Magistrates court and are for less serious offences (although all health and safety cases start in the Magistrate's Court).

Indictable offences

These are held in the Crown court and are for serious offences. These offences do not often apply to health and safety cases unless a fatality or manslaughter is involved (the latter is usually a case for the police, with the enforcing authority helping out in areas requiring specialist knowledge).

The Corporate Manslaughter and Corporate Homicide Act 2007 came into being in April 2008. This piece of legislation is not specifically a part of health and safety law but it will be used for prosecutions of companies and organisations under criminal law, whereby there has been a major breach in the duty of care and serious management failings that have caused the death of a person. The Act could be applied if the breach of the duty of care is seen to have been as a result of a failure by senior management. The Act allows for a prosecution to be taken without having to identify an individual(s) who may be directly responsible. The penalty for a conviction will be a fine imposed upon the company or organisation, and the Court may order that the conviction and fine be publicised. Although it is aimed at the corporate body and cannot be used against particular employees or members of the company/organisation, individuals could still be individually prosecuted under separate health and safety legislation.'

Triable either way offences

These offences are those that can be either summary or indictment – the Magistrates Court may pass it up to the Crown Court if the evidence is there to enable a hearing, but yet they feel their powers of sentencing are not enough for the gravity of the offence.

The Defendant can also choose whether or not to have their case heard at Crown Court, which includes a trial by judge and jury. In these cases, a *mode of trial* hearing is undertaken in the Magistrates Court to decide where it should be held.

Investigations by the Authorities
Evidence
Obtaining evidence to support the alleged breach forms a large part of every investigation. Inspectors will often take away the machine, machine part, relevant paperwork, documents, photographs and other items to prove the breach and the facts of the case. Witness statements are also taken and used in evidence.

The *burden of proof* is on the prosecution to prove that the breach was committed beyond 'reasonable doubt' as opposed to the *balance of probabilities* used in civil cases. Parties are therefore innocent until proven guilty. If, however, the defendant is the only one involved in an incident, and is the only one who knows what happened, then that defendant may have to prove their case (and their innocence) on the balance of probabilities- this is called a *reverse burden*.

Witness statements
Under his or her powers, an Inspector can take a 'Section 20' statement which means that you are required to give the information that they require by law in a written statement. This is quite rare though, and usually this power is only used if someone is not co-operating with the investigation and has refused to give a statement. Normally, a statement is taken with consent and will perhaps contain similar details as ones that you have taken yourself. Statements should not contain *hearsay* which is information with no relevance or that which is inadmissible in court.

161

Hearsay is that which is not relevant to the facts of the case and is omitted because it can often give rise to unreliable evidence because it is based on rumour and not on what directly happened, such as statements like 'He said that she Said' etc. Statements are not to be concerned with opinion, just the facts of what happened. Another concept is that of *res gestae* – which is loosely something that is said during the course of the incident in question that is so closely related to that event that it therefore becomes almost a part of it. So, for example, if Bob and John are walking along together, and Bob falls down a hole, and in the presence of John he says at the time 'This was Bill's fault! I saw him dig the hole!' - then that statement could potentially be used in evidence under res gestae. Res gestae and hearsay are quite complicated, and so I shouldn't worry too much about trying to identify them in the statements that you take within the company during the course of your own investigations.

PACE *interviews*

A PACE interview held by the HSE is an interview under caution similar to that used by the Police, whereby you are warned that anything you say is taken down and can be used in evidence. PACE stands for The Police and Criminal Evidence Act 1984 and this piece of legislation exists to ensure that interviews conducted under caution adhere to certain rules, and describes the powers of the interviewer. Anyone who is believed to have been involved in a criminal offence in terms of health and safety can be invited to attend a PACE interview. The reasons for conducting the interview under caution are that it is then admissible in court as you have been made aware of the consequence of what you say; it can be a way of obtaining evidence against you or the company depending on what you have to say; it can provide further lines of inquiry to follow, and it gives the interviewee a chance to put across their side of the story and any evidence that they have to support this.

A letter will be sent to an individual, or a body corporate (company), inviting you to attend an interview, which is normally held at a local HSE office. You do not have to attend but it might be seen by the courts as un-cooperative behaviour if you do not. You are entitled to have a solicitor with you if you want (although one will not be provided for you at an HSE office, and they are not there to answer the questions put to you). If you are speaking on behalf of a body corporate you may be asked if you have the authority to represent the company if you are not a director or hold another senior position. In this case you may be asked to take a letter from a senior office holder (such as a Company Director) granting you the authority to speak on behalf of the company. A PACE interview is a taped interview, and you are entitled to ask for a transcript of the tape afterwards as you may need to remember what you said should matters proceed to court. Whilst you will be under caution during the length of the interview by the HSE, you are not under arrest and will be free to go at any time.

You may be provided with some disclosure (advanced information about the evidence they may put to you, and the areas of discussion that they may want to talk about), and you are certainly entitled to ask for this beforehand if it is not automatically given to you.

Preparing for court

The Information
An Information gives the details of the particular offence (one offence per information only) for which the defendant is to be summoned to court. It gives details of the alleged breach and the legislation that it is referring to. It is given to enable the defendant to understand why they are being taken to court and to enable them to consider how they will plead in Court. A copy also goes to the Magistrate's Court.

The Summons

The Summons describes the offence and nature of the charge and gives you details of the dates you are required to attend at court. This is served by post or by hand, on an individual or a company as appropriate.

Advance information

This is the information given to you by the prosecution to enable you to prepare for court, and in triable either way cases, to help you decide if you want it to be held in the Magistrate's Court or the Crown Court. If you are not automatically given any Advance Information, you are entitled to request it.

In Court

The Friskies Schedule

This is a list prepared by the prosecution in advance, with facts regarding the case argued between both sides (prosecution and defendant) in advance of the court hearing. Any aggravating and mitigating factors are also given – the prosecution will say if the defendant admitted responsibility, if there are any previous health and safety contacts on the subject such as warnings or Notices given, and on the positive side they will say if you complied with their investigation etc.

Bundle

This is the case file prepared by the prosecution and given to the Court and the defence. It contains the evidence that the prosecution is relying on, statements, the Friskies schedule and other items and is for the Magistrate or Judge to refer to during the hearing.

Newton hearing

A Newton Hearing is often heard if the defendant cannot decide on a plea due to a disagreement with the Friskies schedule and the facts of the case.

The defence may have to bear any costs of a resulting adjournment whilst the facts are argued over.

Guilty pleas

Usually if a defendant pleads guilty in the Magistrates Court they may proceed immediately to sentencing or adjourn for a hearing at a later date. It may lead to a reduced sentence, depending on when a guilty plea was put forward. If the defendant admits guilt at the first opportunity, the sentence may be reduced by up to one third.

If you don't attend court, the hearing may proceed in your absence or adjourn to a later date. They may also proceed with mode of trial in which case you wouldn't be able to have your say on where you want the case to be held. There may be more than one charge against a defendant if multiple offences are alleged. Witnesses may be asked to attend court, and can be cross-examined by both sides. Expert witnesses may be called upon by either side to give specialist evidence.

Abuse of process

Rarely, proceedings are seen to have been conducted so as to be unfair to the defendant, such as that they would not receive a fair trial, or there has been an extensive delay in going to trial. The required attendance in Court should be within a reasonable time of being served the summons.

Double jeopardy

This is the notion that a person cannot be tried again for a crime that they have been convicted or acquitted of previously (although this only applies if we are talking about the same offence – if you commit a similar crime on another day then that is a different matter.)

Sentencing - fines and costs

Fines are applied to deter the defendant from committing the crime again, and to deter others from doing the same thing, along with being a punishment for committing the breach of law in the first place.

165

Under section 33 of the Health and Safety at Work Act 1974, For offences committed since March 2015 the maximum penalty in the magistrates' court is an unlimited fine or imprisonment for a term not exceeding 6 months or both. In the Crown Court, the maximum penalty is an unlimited fine or imprisonment not exceeding two years or both.

The Court may take into account any previous convictions when sentencing, after a guilty plea or when a defendant has been found guilty of a breach. HSE may give guidance to the Court on sentencing, and there is nothing to stop the other side doing the same, in the right situation. The size of the company and its profit and turnover will also be taken into account. The Courts may award costs to the prosecution (HSE Inspectors work out their hourly rate spend working on the prosecution, including the costs of gaining evidence and undertaking the investigation.) Any mitigating factors may be taken into account here too. If you are successful in defending your case, or the case discontinues, you may, in some circumstances, get some of your costs back. Some case law regarding sentencing may be considered, and the Prosecution may give guidelines to the Court, but every case is different with different facts and circumstances, and fines for similar cases may vary for this reason.

Appeals

In certain circumstances, appeals can be made to the Crown or District Courts – with guilty pleas, you may be able to appeal against the sentence received, and with not-guilty pleas, against the conviction or the sentence.

****

Where to get further help and information

- The Health and Safety Executive's (HSE) website - www.hse.gov.uk and the 'info line' that they provide – the telephone number for this can be found on the website

- Your local HSE office or the Environmental Health Department of your Local Authority – details of which can be found in the phone book

- The Institution of Occupational Safety And Health (IOSH) – gives advice and guidance to its members and others on health and safety issues

- The Royal Society for the Prevention of Accidents (ROSPA) – gives health and safety advice on preventing accidents in many different areas of industry, along with information on events and training.

- Relevant magazines and subscriptions which may or may not be industry specific, such as *Health and Safety at Work* magazine that can be ordered from many newsagents and/or bookshops

- Trade and industry associations often produce their own material regarding health and safety which will be specific to your type of work

- For purchasing legislation (much of which can be downloaded free of charge) look at the website of The Office of Public Sector Information (OPSI): www.opsi.gov.uk/legislation or look at the *Acts and Statutory Instruments* pages which can be found on the HSE website home page under the *Legislation* section.

- Information on fire safety can be found on The Chief Fire Officer's Association's website – www.cfoa.org.uk - which gives guidance on many matters associated with fire prevention and the associated legislation.

- The Law Society (www.lawsociety.org.uk) can help to point you in the right direction to finding a solicitor who may deal with specialist areas of business law.
- The Citizens Advice Bureau (www.citizensadvice.org.uk) offers free information and advice on many legal issues.

****

Index

Accident books, 21
Appeals, 166
Approved Code of Practice, 23, 99
Asbestos, 3, 56, 123, 125

Blood-borne viruses, 78

Cancers, 76
Civil claims, 159
Comfort, 38
Company Directors, 11
Competent persons, 18, 36
Confined Spaces, 115
Consulting with Employees, 135
Contractors, 145
COPD (Chronic Obstructive Pulmonary Disease), 77
Corporate Manslaughter and Corporate Homicide Act 2007, 160
COSHH Regulations, 56
Crown Court, 16, 159, 160, 161, 164, 166

Dangerous occurrences, 149, 152
Death, 148
Department for Work and Pensions, 14
Dermatitis, 73
Diseases, 148, 149
DSE assessment, 41

Electricity, 119
Employee Safety Representatives, 135
Employers' Liability Compulsory Insurance, 12, 17
Enforcement, 157

Environmental Health Officers (EHOs, 14
Environmental impact, 21
European Chemicals Agency, 64
European Directives, 24
Expectant Mothers, 100

Financial Conduct Authority (FCA, 12
Fire, 79, 80, 81, 82, 83, 167
First Aid, 84, 86
Floor surfaces, 108
Forklift trucks, 139

H M Revenue and Customs, 10
Hazardous Installations Directorate, 15
Health and Safety at Work etc. Act 1974, 14, 25
Health and Safety Executive (HSE, 14
Health and Safety Training, 37, 144
Home Working, 113

Improvement Notice, 157, 158, 166
Indictable offences, 160
Inspections, 147

Laboratories (HSL, 15
Ladders, 91
Latex, 75
Law Society, 159, 168
Legionnaire's Disease, 98
Local Exhaust Ventilation (LEV, 50
Lone Working, 111

Machinery, 49, 141